The Science Crusade

by

Kevin Daws

Table of Contents

Chapter One - Into the Crusade

"Tom stared at the screen, rubbed his eyes and sighed. There was no way out. The obvious answer was the rope hanging down from the wooden beam just below the ceiling. But the rope kept breaking.

'JUMP' he typed.

'WHERE TO?' came the computer's reply.

'BEAM' typed Tom, hopefully.

'THE BEAM IS TOO HIGH'

Tom looked at the picture on the screen and studied it closely for the umpteenth time. There was a door to the left and in the right-hand corner was a monster grinning rather nastily. A shaft of light was shining in through a hole in the ceiling and a long rope was hanging down into the middle of the room.

'KILL MONSTER' he typed

'DO YOU WANT A FIGHT?' came the reply, and the monster in the corner crept towards the middle of the room.

'NO' typed Tom and he pressed the RETURN key quickly so that his instruction was understood straight away.

Tom sighed again, very loudly. He was getting fed up with the game. It was too difficult.

"What's wrong, Tom? Are you stuck again?"

Tom jumped.

"Go away, Zak," he said rather rudely.

"Don't call me Zak," said Zoe. She peered over Tom's shoulder and looked at the screen, digging her chin into her brother's neck.

"Why don't you use the rope?" she asked.

"Because it'll break. Can't you read?" said Tom, trying to shrug her away.

Zoe could read. In fact, for an eight year old going on nine she could read rather well. Her spelling wasn't too good. She had a habit of doubling up letters when they shouldn't be.

'TOO TOM' she had written on the envelope of his birthday card two weeks before. Tom had laughed at her for that.

She looked at the right corner of the screen where Tom was pointing. There was an odd-looking face and from its mouth flashed the words, 'THE ROPE WILL BREAK.'

"The rope will break," she read out loud.

"What's wrong Tom? Are you stuck again?"

"Any other bright ideas?" asked Tom. "If not, go away."

Tom was secretly hoping that his younger sister would come up with an idea. Even though she got on his nerves he had to admit that she came up with good ideas occasionally. Perhaps he'd overlooked something obvious and she would spot it.

They both looked at the screen before them. There was a door and a rope hanging down from a beam. In the corner was a monster's head with bloodshot eyes and drooling mouth, mocking them with the words 'THE ROPE WILL BREAK.'

"Open the door, open the door!" said Zoe excitedly.

"I just came in through the door," explained Tom, beginning to lose patience, "and I'm trying to get up there."

He pointed to the opening at the top of the room.

"Climb the rope then," suggested Zoe.

"If I climb the rope it will break." Tom really was getting annoyed now. "Watch," he said.

He typed the words 'CLIMB ROPE' and pressed the RETURN key on the keyboard.

The screen went blank for a moment and then flashed up a message in large red letters that said, 'THE ROPE BROKE. YOU GOT EATEN BY THE OGRE OF THE TWINE.'

A gruesome picture of the monster's head appeared on the screen and there were two legs kicking desperately. The monster chewed, the legs were swallowed and a satisfied belch came from inside the monitor.

'YOU SCORED LEVEL 1 ON THE SCIENCE CRUSADE,' was the final message displayed.

"Satisfied?" Tom asked but he did not require an answer.

Zoe said nothing but Tom could see her biting her lip and trying not to laugh.

'PASSWORD' was now flashing on the screen. Tom typed in his secret password quickly so that Zoe couldn't see what he was doing. '*****' was displayed even though he had typed 'GRIZZ.' This was all part of his security system to stop his tiresome sister messing around with his programs. The disc drive whirred and showed its file contents on the screen. Tom selected 'SCIENCE CRUSADE' and before long was back into the game. Eventually he was back in the dimly lit room and the monster, the Ogre of the Twine, was repeating 'THE ROPE WILL BREAK' over and over again.

"Tom! Zoe! Will you come down, please? It's time for lunch," shouted their mother from downstairs.

"Ha, ha, you'll have to start again," said Zoe gleefully, and she rushed out of the room.

Tom typed 'GET OUT' using the keyboard and pressed RETURN. This way he could get back to the exact point in the program without having to start all over again, and it would also stop any nosey kid sisters from interfering. All he had to do after lunch was type 'GET IN' and he would be straight back into the Rope Room ready to outwit the Ogre of the Twine, somehow. He was glad of the break actually. His eyes were beginning to lose focus, and perhaps during lunch he would solve the problem in his mind without thinking about it.

Tom went downstairs and into the kitchen to wash his hands. Zoe was already sitting at the kitchen table and his mother was busy serving up lunch onto three plates.

"Tom got eaten by the Rope Monster," said Zoe, rather too loudly.

"It was your fault, Zak," Tom said, crossly, "and it was the Ogre of the Twine, not the Rope Monster," he corrected.

"It wasn't my fault. You don't let me touch your darling computer, do you Tom? So how could it possibly be my fault?"

"Quiet, both of you," interrupted their mother. "Tom, have you been staring at that computer all morning?"

"Yes, he has," said Zoe. "And he can't get it to work."

"Yes I can. I was doing fine until you came along," said Tom, raising his voice.

"That will do," said their mother. "This afternoon I think it would be a good idea if the pair of you took Grizz out for a walk in the fresh air."

She said this with such authority that Tom knew it would be pointless to argue. Besides, he quite liked taking the family dog out for walks. It was the possibility of being seen with Zoe that worried him a bit. No thirteen year old boy likes to be seen going around with an eight year old girl.

"Wasn't Zak going with you to the library this afternoon," he said, hopefully.

"No, she's going with you and Grizz to Hitchin's Wood. And how many times have I told you not to call her Zak. Her name is Zoe."

Tom had made up the name Zak a while ago. His sister's name was Zoe Alice Kennedy and the initials spelt Zak. Tom thought it suited her rather well because it made her sound like a robot and she was always following him around. "She hasn't got a mind of her own so she must be a robot," he would say.

"She has got a mind of her own," Tom's father would say in his daughter's defence. "She just wants to see what you're up to. So call her Zoe, please."

Tom would usually remember to call her Zoe if his parents were around but sometimes he forgot. Also, he knew that Zoe got annoyed when he called her Zak, and it was one small way in which he could get his own back.

Tom decided to keep quiet for the moment. His mind went back to the computer. He had got it as a birthday present two weeks before and he could not believe his luck as he knew how expensive they were. "It's not for playing games on, of course," his father had said as Tom had excitedly ripped the paper from the box. "It's for writing homework reports and doing calculations."

"Yes, Dad," said Tom, loading up one of the games that came with the package.

"I can get some special paper from work so that your assignments look really professional," his father continued.

"Yes, Dad. Great," said Tom, his eyes glued to the screen as he zapped some alien blobs to pieces, and his father went out of the room.

It wasn't long before Tom got bored with zapping games. They were all the same and not very interesting. He much preferred mystery games where he had to solve a problem to make progress. The latest and best one was The Science Crusade which was fascinating, but also proving very difficult. How was he supposed to get out of the Rope Room?

"Is there anybody in there?" his mother said, waving her hand in front of his face. Tom blushed. He had been in a daydream thinking about the Crusade. Zoe was giggling uncontrollably, her shoulders jumping up and down with the effort of trying to stop herself laughing out loud.

"I was just saying that if you give me your watch I'll get you a new battery. I'm going into town this afternoon."

"OK, Mum," replied Tom. "I'll just go and get it."

When he got back to the kitchen Tom realised that his afternoon had been planned out. His mother was telling Zoe to get her coat and find Grizz's lead so that they could take him for a walk. Grizz was their eight year old spaniel. Tom was only five when Grizz had been introduced to him and apparently he had been terrified by the puppy. He had thought that it was a bear and that was how Grizz had got his name. It was short for Grizzly Bear. Only Tom knew that it was also his secret password into the computer.

"Is the computer switched off, Tom?" his mother said.

"No, but it'll be alright. It hardly uses any electricity."

"That's not the point. You know I don't like to leave things switched if it isn't necessary. Now, go and do it before you go out."

Tom didn't argue. The week before he had left the computer on and gone round to a friend's house. While he was out, his mother had gone into his room, seen the screen flashing away and she had pulled out the plug. When Tom had got back, he found out what she had done. He was furious because he had been in the middle of a mystery game and now it was all lost. There had been quite an argument but in the end Tom had to admit that his mother was right. He would save the game on disc and switch it off.

Tom collected his favourite jacket from the cupboard under the stairs. His mother refused to let him hang it up with the other jackets on the coat rack because it was in such a tattered state. But it was really comfortable and it didn't matter too much if it got any dirtier. It was bright red in colour, if a little stained, and the pockets were full of bits and pieces.

He walked upstairs to his bedroom so that he could save the game on disc. Grizz gave away the fact that the room was not empty. As he opened the door, the little black dog barked and ran over to greet him. He jumped up as high as his chest and, in doing so, knocked over Tom's bedside lamp. Sitting at his desk was Zoe, fingers on the keyboard of his computer and error messages all over the screen.

"Get off!" shouted Tom, angrily at his sister.

"'GETTOFF' she typed and pressed RETURN.

"I was only trying to help," she said, laughing as she did so.

'ERROR. COMMAND NOT KNOWN,' came the computer's reply.

Tom was busy picking up the lamp and trying to push Grizz to one side.

"Get out!" he shrieked.

'GETTOUTT,' typed Zoe and then she pressed RETURN.

'ERROR. COMMAND NOT KNOWN.'

Who said you could get in?" shouted Tom, trying to deal with Grizz who was excited by the whole incident.

'GETTINN,' typed Zoe. Spelling was not one of her strong points.

As she pressed the RETURN key Tom made it across the room, dragging Grizz along with him, and he pressed the ESCAPE key.

The screen went blank for a few seconds, then a picture flashed up. It was a picture of a dimly-lit room but it was not a picture that he had seen before. In the corner of the room was a door and hanging from a hook on the door was a dressing-gown. Also in the room was a bed and on the bed was a lamp, the shade at a strange angle. In the top left-hand corner of the screen flashed the instruction, 'PASSWORD' and at the bottom of the screen was a letter M.

"Now look what you've done," said Tom. He was extremely annoyed.

'GRIZZ,' he typed. '*****' appeared on the screen and he pressed RETURN.

'ILLEGAL PASSWORD . . . M,' came the reply message.

Tom tutted to himself at his mistake. He tried again. '*****,' appeared again as he typed his password. Once again he pressed the RETURN key.

'ILLEGAL PASSWORD . . . M,' came the reply.

It was then that he noticed Zoe gripping his arm tightly. Grizz had gone strangely quiet and there was a damp chill to the room. He shivered and turned his gaze away from the computer screen. He was in a dimly-lit room. A shaft of light was coming in through a hole in the ceiling and a rope was hanging down from a central, wooden beam. Its end lay coiled upon the dusty floor. To his right was a large, oak door and in the corner was a huge head. The head was drooling saliva from its mouth and was piercing him with its stare from bloodshot eyes. Its breath broke the eerie silence with a hiss and stank of rotten eggs.

"The rope will break," it said, and a horrible grin spread across its hideous face.

Chapter Two - The Rope Room

For a moment Tom was completely stunned. What was going on?

"I am the Ogre of the Twine,

And if you don't get out in time,

Tom, Zak and Grizz you will be mine," sang the monster in the corner of the room.

The head was laughing now. It started as a chuckle but before long it was a deep-throated roar that shook the room and sent shivers down the spine of all that heard it.

"What's going on, Tom?" said Zoe in a small, frightened voice. Grizz howled miserably, shrinking under the computer monitor.

"I think we're inside the computer game. This is just like the Rope Room in the Science Crusade and this is where I left the game just before lunch," said Tom, in hushed tones. A moment later he added rather less calmly, "What on earth did you type into the keyboard? I told you to leave things alone."

"It's not my fault, Tom. I typed 'get in'," said Zoe. "I know that's what you have to do."

"Well, that should have been alright," said Tom. "I don't understand this at all. It's creepy. How did you spell the words?"

"I put one word. G, E, T, T, I, N, N," Zoe spelt out, carefully.

"Well, that's why, you fool. You don't spell it like that. Instead of us looking in through the screen, we've somehow ended up inside the computer. We've become part of the game."

"So this should be fun then," said Zoe. "You're good at computer games."

"I've never solved this one though." Tom's voice was shaking. He didn't know how to get past the Ogre of the Twine and he had only managed to get to Level One on this game after hours of trying.

He turned back to the computer screen and, in desperation, he tried his password again. '*****' appeared as he typed. He pressed RETURN.

'ILLEGAL PASSWORD . . . M.'

"I can't get back into the program. My password isn't working. I'll have to try another word but I don't know what to try. It could be anything."

"What's the letter M for?" asked Zoe.

Tom ignored her.

"Quiet, I'm trying to think," he said, impatiently. He was sure that there was an easy way back into his bedroom. Safety was so close yet somehow so far away.

'HELP,' he typed. Sometimes that worked, something had to work.

'****,' appeared on the screen. He pressed RETURN.

'ILLEGAL PASSWORD . . . M.'

There was no way that he could guess the password. This password belonged to the Crusade and the monsters in the Crusade. If he only knew the password he was sure that he could get out of the program by typing GET OUT or GETTOUTT. He had a feeling that it was the misspelt command from Zoe that had caused all the problems in the first place.

The Ogre was still singing and laughing in the corner and creeping ever closer towards them. Tom turned back to the computer and frantically tried all kinds of possible passwords.

'KEYBOARD.' 'THOMAS.' 'MONSTER.' Anything he could think of. Every time came the same reply, 'ILLEGAL PASSWORD . . . M.'

His concentration was broken by the sudden screeching of the slimy head inching towards them.

"A guess at letters can't be made,

You have to solve the Science Crusade," it sang, repeating the lines over and over again.

"Tom, let's get out of here now," shouted Zoe above the Ogre's song. She had her back to the wall and Tom could see that she was very close to panic.

"Calm down a minute, Zak. Look, we've got to solve the Crusade. Listen to the monster. It's the only way out."

"But how?" said Zoe, tears welling up in her eyes.

"We've got to get up there." Tom pointed to the hole in the ceiling above them.

"But how can we get up there, Tom? It's too high to jump and the rope will break if we pull it."

"We must use the rope somehow," said Tom, logically. "It must be there for a reason."

The rope was tied to a wooden beam directly under the hole in the ceiling and about four metres above the floor. Tom pulled on it very gently.

"The rope will break, the rope will break," screamed the head, its voice spitting and screeching the words with excitement.

Tom had to admit to himself that the rope did not feel very strong. He was also very aware of what had happened when he had played the game from the safety of his bedroom. The rope had snapped, he'd been trapped in the Rope Room and the Ogre of the Twine had slithered over and devoured him. End of game. Well, that was alright if you were playing the game on computer but not when you were playing the game for real. He couldn't afford to make a single mistake. He stopped tugging on the rope.

"Mum! Mum!" Zoe was shouting at the top of her voice. Grizz obviously sensed the panic in her voice and started to bark. The Ogre laughed loudly, making more noise than the pair of them put together.

"She cannot hear, she cannot hear,

She's out there and you're in here," it squealed, veins throbbing in its forehead.

Meanwhile, Tom was emptying the pockets of his old, red jacket. He was searching for something, an idea, anything. There was months' worth of rubbish in there including a magnifying glass, an old conker on a string, a balloon, a compass, some bubble gum but nothing that helped him solve the problem.

"Tom, what are you doing? We've got to get out of here and you're looking for bubble gum," yelled Zoe, and with that she dashed past Tom towards the door on the other side of the room.

"Come back, Zak," called Tom. "It's no good. We've got to go upwards."

But his words were not heard. Zoe pulled the large, wooden door open and dashed through it. Grizz was hot on her heels and the door slammed behind them.

Tom had no option but to follow. The smell of the Ogre's fetid breath was overpowering anyway but he thought it best that they should all stay together. He ran over to the door, yanked it open and ran out into the sunshine.

"You may be out, you may be out,

But soon you will begin to doubt,

From here you took the best way out," squealed the monster.

The door slammed on the Ogre of the Twine.

It was a relief to be out in the fresh air again.

"Where are we now?" said Zoe, who was leaning against the wall beside the door.

"It must be part of Hitchin's Wood, I think," said Tom, looking around and trying to spot a familiar landmark. He knew that he'd been to the place before but he couldn't quite work out where it was or when he'd been there. There were trees in all directions except for a long, stone wall and the door from which they had just escaped. Grizz was busy burrowing at the base of the wall under some ferns that were growing there.

"This isn't Hitchin's Wood, I'm sure," said Zoe. "It isn't as muddy as this."

"Where else can we be, Zak," said Tom. "This is where we planned to go this afternoon."

"But what about the room and the monster and the rope . . . ?"

"That was nothing," said Tom, interrupting her and speaking with renewed confidence. "We imagined the whole thing. If you spend too long in front of the monitor it makes your eyes go funny. It was scary though for a minute. I was nearly sure that it was real."

"But I saw it all as well, Tom," said Zoe. "And I only spent five minutes playing the game and that was this morning. Anyway, we can't both imagine the same thing."

"Why, what have you been imagining, Zak? I'm quite sane. I didn't see a thing," said Tom, mocking his sister, and he strode off along the boggy pathway through the woods.

"Come on, Grizz, Let's go," he called. He whistled to his dog to come, happy in the sunshine even if he was a little confused by what had gone before.

Grizz did not appear. He was still burrowing at the bottom of the wall.

"Grizz!" shouted Tom, sternly, turning round and making his way back along the muddy path. His trainers were filthy now so he hardly bothered finding the best route. When he got to the wall he grabbed the spaniel by the hind leg and pulled. Grizz came out backwards and in his mouth was the corner of a computer keyboard.

"Put it down, boy," said Tom, and the dog obeyed. Tom did not understand what was going on. How could there be a computer out here in the woods? He got down on his hands and knees and crawled under the ferns. He followed the wire of the keyboard with his hand and eventually came across a computer screen built into the stone wall. On the screen was a picture of his bedroom. There was the broken lamp lying on the bed and his dressing-gown on the bedroom door. 'PASSWORD,' flashed on and off in the top lefthand corner of the screen.

Suddenly, he knew why he recognised the wood. He hadn't wanted to believe it before so had blocked the thought from his mind. This was the scene before the Rope Room in the Crusade. It was called Marsh Wood and it was the first level in the computer game. He groaned, loudly.

Meanwhile, Zoe had crawled under the fern and was next to him staring at the screen.

"What's the matter, Tom?" she said. "Have you hurt yourself?"

"We're still inside the computer, Zak. Look, there's my bedroom as we left it."

"Where's the M?" said Zoe, unexpectedly.

"What M?" said Tom.

"There was an M before," said Zoe, but the words hardly made any sound as she spoke.

They crawled out from under the fern.

"I know where we are," said Tom. "This is Marsh Wood. I recognise it from the game. It's the level before the Rope Room and if we stay here the ground will swallow us up."

"What do you mean, Tom?" asked Zoe.

"In the Crusade, Marsh Wood gets more and more boggy. Eventually we'll sink into the ground."

"Can't we stand on a rock or something?" said Zoe.

"Not really," replied Tom. "Eventually the rock will sink with us standing on top of it."

"So where do we go from here?" said Zoe. "And don't say back in there."

"Back in there," said Tom, nodding towards the door in the wall. "It's the only exit from this level. I tried everything else when I was playing the Crusade before. I tried standing on rocks. I tried climbing trees. I tried crawling into caves. And every time I got swallowed up by the sinking mud in the marsh and lost the game. The only way out is through that door."

"Well, I'm staying here," said Zoe.

Before he could reason with Zoe, Tom's attention was distracted by the sound of Grizz who was whining pitifully. Looking round he saw that the poor animal had sunk into the ground up to his belly. Tom pulled him out of the mire. It was quite an effort but eventually his feet came loose with a sucking sound.

"Why is Grizz sinking more than us?" said Zoe, noticing that her own shoes were now below the surface and covered with foul-smelling mud.

"It's because his feet are so small," Tom explained. "The smaller your feet are, the more pressure there is on the ground. He doesn't weigh very much but, even so, he must be pushing down with more pressure than we are."

"I don't get it," said Zoe. "Grizz is really light compared to you or me."

"It's the same idea as Eskimos wearing snow shoes. You know, those things that look like tennis rackets that they wear on their feet. The snow shoes make their feet bigger so the pressure on the ground is less. That way, they don't sink into the snow."

"So I'll sink next because I've got little feet," said Zoe, not looking too happy with her reasoning.

"Probably," said Tom. "Even though I'm a bit heavier than you, I have got much bigger feet in comparison. Anyway, I think we should both sit down to spread the weight."

They both sat down on the muddy ground, leaning against the wall. Tom was right about spreading the weight. They hardly sank into the ground at all. Tom held Grizz on his lap, not wanting to risk drowning the little dog in a sea of mud.

They both sat down, leaning against the wall

Tom was seriously worried. He honestly did not know what to do. Absentmindedly, he fiddled with the items in his pockets whilst he thought.

If only they had a stronger rope. Then they could climb up to the beam in the Rope Room and get out through the hole in the ceiling. His hand grabbed the conker in his pocket and he pulled it out. He twirled the string around his finger. Round and round it went, first one way and then the other.

This was his champion conker, undefeated after seventeen fights and perhaps never to fight again.

"Do you remember when I saved your conker last week?" said Zoe.

"Yes, I was just thinking about that myself," said Tom.

"I risked my life for you and that conker," said Zoe. "So now you can save mine."

Tom remembered the incident well. It was a championship fight and the strings had got twisted. His opponent had pulled hard and the string on Tom's prize conker had snapped. The conker had hit the ground and was just about to be stamped on when Zoe, who had been watching the contest with interest, had dived on top of it and saved it just in time. It was hardly life threatening but Tom had been grateful nevertheless.

"The string broke, didn't it?" said Zoe. "What did you do to make sure that it didn't break again?"

"I doubled up the string," said Tom. "That made it twice as strong because each part of the string could take half of the pulling force."

There was silence for a moment as Tom put his mind back to their immediate problems.

"That's it!" he cried. "That's how we can use the rope to climb up to the beam. There's enough rope, I'm sure."

He leapt to his feet as best he could. The ground had become very soft and he had sunk a few centimetres into it. As he stood up his right leg sank down as far as the knee.

"We'll have to crawl on our stomachs," he said. "The ground has become too soft to walk on and if we wait much longer we'll sink completely. Quick, let's go."

"Where?" said Zoe, hesitantly.

"Back to the Rope Room. Come on, we haven't time to talk about it," said Tom, sternly.

Zoe didn't argue. They crawled over to the door in the wall. It was not easy because the ground was very soggy and they had to spread their weight as much as possible to stop themselves from sinking. Grizz stood on Tom's back, obviously sensing that he would not survive in Marsh Wood otherwise.

Pushing open the door, they went back into the chill of the Rope Room. The door slammed shut behind them as they got to their feet and there was a click as the locking mechanism jammed into place. Tom tried the door and it was locked. Their time in Marsh Wood had apparently come to an end and they could not return there even if they wanted to.

"Look, the M's back," said Zoe, looking at the computer monitor inside the room but before Tom could see they were interrupted by the Ogre of the Twine.

"I thought you'd sink, I thought you'd sink,

But now you've made it from the brink,

It must be time to eat, I think," bellowed the slimy head in the corner, appearing even more threatening than before.

Tom rushed to the middle of the room and picked up the coil of rope on the floor.

"The rope will break, the rope will break," squealed the monster.

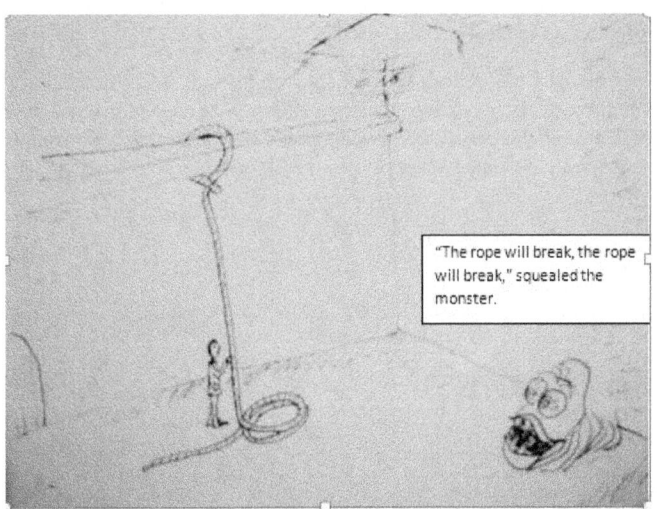

Tom threw the end of the rope upwards, trying to loop it over the top of the beam. It wasn't an easy thing to do and he missed. He picked up the end, made a smaller coil and threw again. This time he was more accurate and managed to get it over. He jumped up to grab the loose end and pulled it gently downwards. By doing that he now had three thicknesses of rope made up from one loop doubled back over the beam and one loose end. He twisted the three lengths together and tied them securely using the conker string. He now had a rope three times as strong as before and hopefully strong enough to take all of his weight. He pulled down on the thicker rope that he had made. At first he pulled gently but he had much more confidence in it now so he pulled a little harder. He noticed that the Ogre had gone strangely quiet. He pulled down with all his weight hardly daring to breathe in case the rope snapped. The rope held.

Now the three of them had to climb up the rope and onto the wooden beam. That was fairly easy for him, but what about Zak and Grizz?

"Zak, have you got Grizz's lead," he said, urgently.

"It's here," said Zoe, pulling it out of her pocket. Tom made a coil with the lead and passed it under the front legs of the little dog and around his belly. He then tied Grizz to the rope hanging from the beam. He then turned to Zoe.

"You've got to climb the rope," he said. "I'll push you as far as I can and then you'll have to haul yourself up. You've got to do it Zak."

A little to Tom's surprise, Zoe was able to climb the rope quite easily and without any help from her big brother. In fact, if anyone struggled to climb the rope, it was Tom. Anyway, he heaved himself up and then lifted Grizz from down below like a rescue helicopter lifting someone from a shipwreck. When Grizz was standing on the beam, legs shaking a bit, Tom untied the rope and let it fall back down into the Rope Room. The Ogre grabbed the rope and pulled hard. It snapped with a twang and then the monster shrank back into the corner of the room, for once with nothing to shriek or sing about.

Tom peered out through the hole in the ceiling. He pushed Grizz through, then Zak and finally hauled himself into a courtyard about twenty metres square. The courtyard was empty but for a couple of trees. Between the trees was a huge spider's web, and, sitting motionless high up in one of the trees, was the biggest spider they had ever seen.

Chapter Three - The Spider's Web

"Quiet," whispered Tom to the others. "Let's get into the shade behind those rocks in the far corner and think about what we're going to do next."

They crept over to the hiding spot and sat down facing away from the web. Grizz huddled in between them.

"So how do we get out of here?" said Zoe before Tom had even thought about their next move. "Things are just going from bad to worse."

"Just be quiet, Zak, for once in your life."

Tom peered over the rocks and studied the latest situation. They were sitting in one corner of a dusty courtyard surrounded on all four sides by a high, stone wall. In the middle of the wall opposite was a wooden door just like the one leading into the Rope Room. Above the door was a sign with some writing on it but the writing was too small to read from where they were.

As Tom had already noticed, there were also two trees in the courtyard which had long, smooth trunks before the first branches. There were a few dead-looking leaves on the branches and Tom guessed that they were probably beech trees although he couldn't be sure. Between the trees was a perfect spider's web. It glistened in the sunlight and shimmered as the wind gently moved the branches that it was connected to. And high up in one of the trees was the owner of the web. It was an enormous, long-legged, hairy spider. Tom sat back down and reported his findings to Zoe.

"There's a door opposite and I guess that's how we get out. The only problem is that there is a member of the arachnid family in the way."

Tom did not want to mention the word spider just yet because he was sure that Zoe was terrified of them, just like all girls.

To his astonishment, Zoe said casually, "Oh, you mean the spider up in the tree. We'll get past that easily."

"Have you seen it properly?" said Tom. "That pathetic spider is about twenty times normal size. Do you know what spiders do to their prey?"

"No, I don't," said Zoe, "but Mum says that they can't hurt you so there's nothing to be scared of."

"Well, Mum has never seen a spider as big as this one, has she? Spiders inject liquids into their food that dissolve it away. Then they suck up the soup that they've made. Sometimes, they stun their prey with a bit of poison to stop it wriggling around. That thing out there is big enough to find us very tasty."

"It won't even know that we're here," said Zoe, peering over the rock to assess the situation for herself. "All we have to do is creep around the edge of the courtyard and over to the door. As long as we don't touch the web we'll be alright."

Tom had to agree that she was probably right. He had watched spiders before. They were quite lazy creatures really once they had spun the web. They just sat there waiting. If an insect got caught in the web it would struggle to get free and that would alert the spider that dinner had arrived. He had never seen a spider leave the web and go out hunting for food. As long as they didn't get caught up in the web they would be fine.

"Right then," he said. "Let's walk around the perimeter of the courtyard with our backs against the wall. If the spider moves, we stand still. I don't think it can see very well and only then if things are moving."

They crept around the edge of the square. Tom led the way, checking carefully that there were no hidden lines of silk that the spider had spun as a trap. There didn't seem to be any. About halfway round to the door there was an alcove in the wall with space enough for the three of them to squeeze into. Inside the alcove was a computer keyboard and monitor. Tom knew that the screen would be displaying his bedroom and hardly gave it a second glance. Zoe seemed rather more excited by what she saw.

"Look, Tom," she whispered.

Tom looked at the screen. There was his bedroom as he thought and in the top left-hand corner was the flashing sign saying 'PASSWORD.'

"Down there," said Zoe, pointing to the bottom of the screen. There were two letters now, an M and an E.

"ME," she said. "What does it mean, Tom? Why is there another letter? First we had an M, then we had nothing and now we've got ME."

"It's a clue," said Tom. "It's a clue to help us to get out of the Crusade. But I'm not sure what it means."

"Perhaps it's the password," said Zoe, excitedly. Before Tom could reply she typed 'ME' followed by RETURN.

'ILLEGAL PASSWORD . . . M E.' flashed the reply message.

"Perhaps it means me not ME," she said, making no sense to Tom whatsoever.

She typed her name 'ZOE' and then pressed RETURN.

'ILLEGAL PASSWORD . . . M E.'

"Or perhaps it's me," said Tom, now seeing what Zoe had meant. "After all, it is my computer."

For the next few minutes they tried writing their names into the computer to see if they could get the password. They tried first names, second names, surnames, first names and second names, surname then first name, all names. Every time they got the same frustrating error message, 'ILLEGAL PASSWORD . . . M E.'

"Maybe it's the spider. Perhaps the spider owns the Crusade and the password is its name," said Zoe.

"What's the spider's name then, Zak?" said Tom, beginning to get irritated. He was annoyed because he knew that 'M E' was a clue and he wasn't clever enough to know what it meant.

"Do you think it's Incy-Wincy. Why don't you go and ask. Please Mr. Spider, don't bite me, but can you tell me your name?" he mocked. It wasn't a very nice way to talk to Zoe but he was frustrated.

"Come on, let's get to the door," he said in a normal voice, changing the subject quickly in case Zoe decided to carry on the discussion and disturb the spider in its lair. They left the alcove and sidled on towards the wooden door.

Just then, the wind rustled the leaves of the towering trees in the courtyard. One or two crisp, brown leaves fluttered downwards and got caught in the web. The web jerked for a moment and in an instant the spider was alerted and on its way down its fishing thread. It obviously thought that some unlucky prey had got caught in the sticky silk and that dinner had arrived.

Tom, Zoe and Grizz stood dead still, hardly daring to breathe as the spider suddenly appeared in the middle of the web. However, it didn't stay very long. Having discovered the reason for the disturbance of the web, it returned to its vantage point high in the treetop. With much relief the three continued on their way, much quieter than before.

Having reached the door, Zoe carefully turned the handle and pulled. The door held fast. She tried pushing but that didn't move it either.

"It's locked," she whispered.

Tom was not surprised to hear the news. It would have been far too easy to get out of this particular section of the Crusade if all you had to do was open a door quietly. He was busy reading the notice above the door. It said, 'USE THE MAGNETIC KEY. THEN HEAD DUE NORTH BY FOLLOWING YOUR NOSE.'

"What magnetic key?" he mumbled to himself.

"We've got to think, Zak," he said, more loudly so that his sister could hear. "Let's go back behind those rocks so that we can't be seen. We've got to find the magnetic key so that we can unlock the door."

They crept back to their hideaway.

"What's the magnetic key?" asked Zoe when they had settled themselves down out of sight of the arachnid.

"I don't know," replied Tom, "but the answer must be close by otherwise the Crusade is impossible."

For some minutes they sat in silence. Grizz had never been so well behaved but he was not stupid and must have sensed the danger that they were in.

"It's the password! That's what the clue means," exclaimed Zoe, suddenly. "How do you spell magnetic?"

"M, A, G, N, E, T, I, C," said Tom, a little puzzled by the question. But then he realised what his sister was getting at.

"Of course," he said. "M and E are two of the letters of the word magnetic, and that's the key to the password. The magnetic key! We're out of here. Let's go back to the computer and type it in."

At a quick snail's pace they edged back to the alcove and the computer terminal. The PASSWORD message continued to flash and the letters M and E were still in the corner of the screen.

"Here goes," said Tom and he typed the word 'MAGNETIC.'

'*********' came up on the screen. Zoe picked up Grizz and Tom held Zoe's hand as he pressed RETURN.

'ILLEGAL PASSWORD . . . M E,' came up on the monitor. Tom's heart sank but he didn't give up immediately. He tried 'MAGNET', 'MAGNETISM,' 'MAGNETISE,' 'MAGNETISED,' and even 'MAGNETISATION' although he knew it wasn't a proper word. With each attempt he lost more and more hope and he slumped down onto his haunches with his back against the wall.

"Why didn't it work, Tom?" asked Zoe.

"I don't know, Zak. It must mean something else. Use the magnetic key, use the magnetic key, use the magnetic key. What does it mean?" he muttered to himself.

"Perhaps there was a key in the Rope Room," said Zoe, "hanging on the beam or something. Or maybe the monster had it."

"I don't think so," said Tom. "We solved the problem in the Rope Room and now we've got to solve this."

"Well I think you should go and look," said Zoe.

"OK," said Tom, sighing loudly.

He told Zoe to take Grizz back to their position of safety behind the rocks and then he crawled over towards the hole in the middle of the courtyard. His movements were very, very slow as the web was only just above his head. Just then there was another puff of wind and more autumnal leaves fluttered downwards. One or two got caught in the web and immediately the spider came shooting down to investigate.

Tom stayed absolutely still hardly daring to look, but he did so all the same. The spider was a magnificent specimen. It looked big enough to swallow a dog whole although Tom knew that was not the way that it fed itself. He could imagine it sucking out his juices and then wrapping his bones in a parcel of silk and tossing the bundle away like a used napkin. He shuddered uncontrollably. The web was obviously not designed for insects because the threads were too far apart to catch anything that small. This web was surely meant for larger prey.

The spider returned to its den and gave no sign that it had seen or smelt anything unusual. Tom watched it go and it was then that he noticed the keys. There were three of them stuck near the centre of the web. From a distance they had looked like leaves which is why he hadn't noticed them before. But this must be the way out of the spider's lair.

There was a massive problem though. He guessed that he would only just be able to reach the keys if he stretched up and they were too far apart to grab them all at once. As soon as he grabbed one of them the spider would come zooming down to investigate. There would not be time to grab more than one key and get away to the door, unlock it and escape before the spider caught them. But suppose he chose the wrong key. The keys were too high for Zoe to reach and Grizz couldn't help at all. It seemed as though their lives depended on the one chance in three in selecting the right key. He crawled back over to the others to report his discovery.

"Was there a key?" said Zoe.

"No, but listen, I think I've worked out what we've got to do to get out of here."

Tom pointed out the three keys to Zoe. From where they were sitting they could just about make out the shapes hanging from threads near the centre of the web.

"So here's the plan," said Tom. "You and Grizz will creep over to the door. When you've got there I'll grab one of the keys and we'll see if it unlocks the door."

"What if it doesn't?" said Zoe.

"We get caught by the spider I suppose, and we lose the game. I've got one chance in three of picking the right one but what else can we do?" Tom said, hoping that Zoe would come up with some sort of idea.

"But it's easy, Tom," she cried. "All you have to do is pick the magnetic key. That's the one that opens the door, just as it says on the sign."

"I suppose you're right, Zak," said Tom. "But how do I pick the magnetic one? They all looked exactly the same to me."

"But there has to be an answer, Tom, doesn't there? Otherwise it wouldn't say on the sign."

Tom knew that Zoe was right. There had to be a way of picking the correct key first time. It made sense that there had to be an answer. This was the Science Crusade, not a game of luck.

Tom sat down to think. He knew that very few metals were magnetic. The commonest one was iron but there was also nickel and a couple of other ones that he couldn't remember. And there were also a few magnetic alloys; mixtures of metals that contained at least one magnetic metal. But it was impossible to tell if a metal was magnetic just by looking at it. You had to test it out with a magnet. If the magnet stuck to the metal, then the metal was magnetic. But, right now, he didn't have a magnet.

But wait, yes, he did have a magnet. The compass needle was a magnet. That was why it lined up with the Earth's magnetism. And he had his compass with him in the pocket of his jacket. All they had to do was to put the compass close to each of the keys in turn. The key that made the compass needle move away from North was the magnetic key.

Barely able to hide his excitement, Tom whispered his plan to Zoe.

"Zak, I've worked it out. You see my compass?"

"Yes," she said.

"Well, at the moment it's pointing North. Look, it's pointing at the door and it does that even if I twist it around. The needle always points to the door." He demonstrated the effect so that Zoe could see what he meant.

"Why does it point North?" said Zoe.

"Never mind that at the moment. What you've got to do is place the compass next to each of the keys. If the needle points to the key then it must be magnetic and we can use it to unlock the door."

"Why me, Tom? Why can't you do it? I don't want to go anywhere near the spider," said Zoe, her voice trembling.

"The keys are above my head," said Tom. "The compass needs to be horizontal and upright when it's used, otherwise the needle gets stuck. I won't be able to see the needle if I'm holding it above my head."

"Nor will I then, will I? You're much taller than me," said Zoe.

"Yes, I know," said Tom, "and that's why you've got to sit on my shoulders as I walk past each key. You need to put the compass up close to each key and, when we get back, tell me which key moves the needle. OK?"

"Not OK," said Zoe, firmly. "What if the spider comes down? It'll get me."

"It won't come down though," said Tom with more confidence than he really felt inside. "The spider will only come down if the web moves. And you don't need to touch the web at all. You don't even need to touch the keys because magnetism spreads through the air. Just put the compass about a centimetre away from each key and don't make a sound until I bring you back to here."

"Alright then," said Zoe, swallowing hard.

Tom could tell that Zoe was absolutely terrified. He put his arm around her shoulders.

"Well done, Zak," he said, a little awkwardly. "You're very brave."

"I don't feel very brave," said Zoe.

"Well I think you are," said Tom. "If you're scared and you still try and do something, that's being brave."

With Zoe perched upon his shoulders, Tom walked very slowly and rather unsteadily towards the spider's web.

They decided to put their plan into action. With Zoe perched upon his shoulders, Tom walked very slowly and rather unsteadily towards the spider's web. Zoe held the compass in one hand and grabbed Tom's collar with the other, trying hard to keep her balance. Tom walked from one side of the web to the other, passing underneath each of the keys and pausing while Zoe tested each one for magnetism. Zoe was not heavy but it was difficult to keep steady especially when he felt so nervous. One false move and they could brush against the web and the evil-looking spider would be alerted.

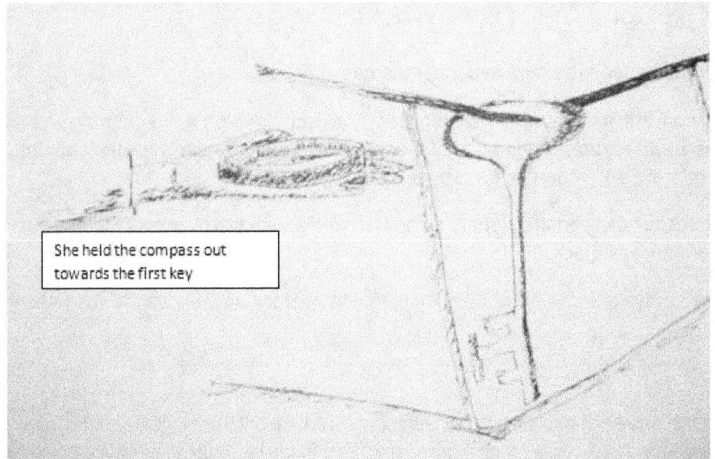

She held the compass out towards the first key

Anyway, assuming that Zoe had tested the keys, Tom staggered back to the hideaway behind the rocks. He lowered Zoe gently to the ground.

"Which key was it, Zak?" he demanded, breathlessly.

"It was the middle one . . "

"Right, let's go. You take Grizz around to the door. When you get there I'll grab the middle key and come and unlock the door. Remember, we'll have to be quick because as soon as I take the key, the spider will be after us. Off you go."

"Wait, Tom. You didn't let me finish. It was the middle key and the one on the left. They both made the compass turn."

"I knew it," said Tom, looking skywards. "Can't you do anything right? Did you hold it steady when it was next to the keys? Did you . . ?"

"Yes, I did," protested Zoe. "I did everything you told me. The one on the right didn't move the compass needle but the other two did." Tears were welling up in Zoe's eyes but she didn't cry.

Tom could see that his sister was telling the truth and he felt bad about the way that he had reacted. Once again, he was blaming her because he had no answer to the problem.

"We'll just have to choose from the two keys," he said. "At least we've narrowed the chances down. Before it was one chance in two and now it's one chance in two. It's still a massive risk though."

Tom was utterly dejected. It was so unfair of the Crusade. If you guessed wrong you could start again if you were on the computer. But if you guessed wrong when you were in the computer you ended up as spider food.

Zoe had been fiddling with the compass while Tom was feeling sorry for himself. As she span it around, the red end of the needle kept pointing at the exit door across the courtyard.

"Why does it always point the same way, Tom?" she said, breaking the silence.

Tom forgot his troubles for a moment to ponder the question.

"It's because of the Earth's magnetism," he replied. "The Earth is like a giant magnet. The North Pole of the Earth attracts the red end of the compass needle which is also a magnet. The South Pole of the Earth repels or pushes away the red end of the compass needle."

He was pointing to the coloured ends of the compass needle as he explained. One end was red and the other end was black.

"So the compass always points North," he said, happy to be back on speaking terms with Zoe.

"Oh, I see," said Zoe.

"That's it! That's the answer," exclaimed Tom, leaping up with excitement. "One of the keys is a magnet and one of the keys is made from a magnetic material. That's how we can tell them apart. And we take the one that is the magnetic. Use the magnetic key. Don't you see?"

Zoe nodded, but her expression did not show that she understood at all.

"We've got to do it again," said Tom, hardly pausing for breath. "At least, we've got to do it again for the middle key and the one on the left. But this time you've got to put the compass close to the top of the key and then move it down to the bottom of the key."

"Why?" said Zoe.

"One of the keys is a magnet. When you move the compass along it, the red end will attract to the North end of the key, but it will repel from the South end of the key. So the compass needle will spin around as you move it down the key."

"But won't both keys do that, like before?" said Zoe.

"No, they won't," said Tom. "The other key is made from a magnetic material and because the compass is a magnet, the needle gets attracted. But the same end will point to the key all the way down. That's how we tell them apart."

"So what do I do again?" said Zoe, nervously.

"Find out which key twists the compass needle around completely. You move the compass from top to bottom. Either the middle key or the one on the left will twist it. You need to find out which key it is."

"What about the key on the right?" said Zoe.

"We know that one is definitely wrong because it didn't affect the compass at all. There's no need to test that one. Do you understand, Zak? It's important that you get this right."

Zoe nodded. "I understand what I have to do," she said. "I'm just not exactly sure why."

Zoe clambered back onto Tom's shoulders and, once again, they quietly approached the spider's web. Tom walked beneath the two keys and then returned to the rocks.

"What did you see this time?" he said.

Zoe explained what had happened, her voice shrill with excitement.

"With the first key," she said, "the red end of the compass pointed to the key all the way down. But with the other key, it did just what you said. The red end pointed to the loop of the key, but as I moved the compass downwards, the needle span around so that the black end pointed to teeth of the key."

"Well done, Zak. I knew you could do it. The middle key is the magnetic one. Let's get out of here."

Tom was thrilled that his idea had worked. The Crusade was fair and it could be solved if you thought about it.

"Take Grizz around to the door. Go around the edge of the courtyard well away from the web. When you get there I'll grab the magnetic key and run over as fast as I can. I'll unlock the door and we rush out. We won't have much time because the spider will be after me as soon as I disturb the web."

Tom's heart was pounding as he waited for his chance to escape. The plan could still go wrong. Suppose he couldn't pull the key away from the sticky threads of the web. Suppose he dropped the key. Suppose he wasn't fast enough. Suppose it was the wrong key after all.

He saw that Zoe had made it to the door on the opposite side of the courtyard. She waved to him, he waved back and then he drew in a long breath to try and calm himself down. He crept up to the web and stood underneath the middle key - the magnetic key. He looked up towards the spider. It was motionless high up in the tree. He stretched up, gulped noisily and grabbed the key. It seemed to come away quite easily in his hand. He rushed across the courtyard not daring to look behind him, placed the key in the lock and turned it. It worked. Tom pulled the door open and pushed Zoe and Grizz through it. As he dashed through himself he heard a chuckling noise and he risked a look over his shoulder. He saw that the spider was still high up in the tree, just as before. The door slammed shut behind him and, as it did so, the lights went out and the piercing howl of a wolf came out of the darkness.

Chapter Four - The Pointers

Stand still and let your eyes adjust to the dark," said Tom. It had been quite bright in the spider's lair and now there was very little light at all. In fact, at first they could not see a thing but after a while they began to pick out points of light in the sky. The stars were shining brightly but there was no other light at all.

Tom could make out the horizon because the ground was completely dark and the sky was filled with stars but otherwise it was difficult to make out anything. He could see the silhouette of his hand when he held it up against the sky, but nothing else. And all the time the howling of distant wolves made the hair on the back of his neck stand up on end.

"I'm scared," said Zoe.

"So am I," said Tom. "But try to stay calm. We won't get anywhere by panicking."

Grizz was scratching around at the base of the stone wall and he drew their attention to another source of light. Set into the bottom of the wall was a computer monitor. It was covered with a black curtain which Grizz had pulled open with his teeth, revealing the pale green glow from the screen. Tom stumbled over to the computer, tripping on some rocks in his haste.

"Zak, walk towards my voice," he said, urgently.

Seconds later Zoe came crashing headlong into him.

"And mind the rocks," Tom said, giggling.

They looked at the picture on the computer screen. 'PASSWORD,' flashed in the top corner of his bedroom. As they looked, the bedroom door opened and their mother walked into the room.

"Tom, Zoe, Grizz," they heard her call in a distant voice, and then in an even quieter voice she said, "That's funny; I never heard them go out."

Tom immediately closed the black curtain, took off his jacket and covered the curtain with it.

"What are you doing, Tom?" said Zoe. "Mum was on the screen. She might be able to hear us."

"She won't," whispered Tom. "And don't talk. She might hear us."

"You're not making any sense," said Zoe. "Mum! Help!" she shouted, trying to rip Tom's jacket away from the screen.

Tom rolled her away from the computer and put his hand over her mouth.

"Be quiet, Zak," he whispered at her.

For a few seconds he let her struggle but then he felt her relax a little. The wolves howled a bit louder, obviously closer, and Tom felt Zoe tense up again. Without releasing his grip he said quietly,

"I don't want Mum to think that the computer is on in my bedroom. Last week she switched it off when I was out. If she does that now, we'll be lost inside the computer forever."

He took his hand away from Zoe's mouth and she drew in a huge gulp of air.

"But Mum would see us on the screen and help us," she said, breathlessly.

"I don't think so," said Tom. "If she saw little figures on the screen she would hardly think it was us. She would think it was some sort of computer game and switch it off. In some ways we're lucky it's so dark at the moment. The screen will appear switched off already."

"But I'm scared again, Tom. I don't like the dark and that howling is getting worse."

Tom put his arm around Zoe's shoulders and Grizz clambered between them. For a moment he felt secure, but the barking and howling was getting nearer and he knew he had to act fast.

Carefully, he peered behind the black curtain. The bedroom door was closed and his mother had left the room. He breathed a sigh of relief. At least they hadn't been eliminated from the game by his mother switching the computer off. He noticed that the lamp had disappeared from the bed but otherwise everything was as it had been before. At the bottom of the screen were the letters M, E and H.

"We've got another clue," said Tom, into the darkness.

Zoe joined him moments later.

"What do you think, Tom?" she said.

'H E M,' typed Tom, then he pressed RETURN.

'ILLEGAL PASSWORD . . . M E H,' came the reply message.

'MESH' gave the same result.

"Home! It's home," said Zoe, with excitement.

"You could be right, Zak," said Tom, typing in the word at the same time.

But the same error message appeared.

"It's no good," Tom said with disappointment. The word 'HOME' seemed to make sense and he had been quite hopeful that it was right.

"What next then, Zak?" he said, letting the curtain drop back into place and putting his jacket back on.

"We go North," said Zoe, without hesitation. "That's what it said on the door, remember? 'Head North and follow your nose.' Quick, get your compass out."

Zoe was turning out to be quite useful, thought Tom to himself. He had forgotten about the instruction on the door and there was no way that he could have solved the other problems by himself. He grinned, knowing that Zoe could not see his face.

"Yes, I was just about to do that," he said. "But unfortunately, I can't see the compass. It's too dark and the needle doesn't glow. It's not fluorescent."

"Well, hold it up to the computer screen then," said Zoe.

Tom did as he was told. In the glow of the screen he could just make out the small needle.

"I can only just see it," said Tom.

"Well, never mind," said Zoe. "I've got another plan. The door is here," she said, knocking on it with her fist.

"Yes," said Tom. "So . . .?"

"I remember that the door was North when we were in the courtyard. So if we walk straight ahead from here, we should be alright. And that's why the sign said, 'Head North. Follow your nose.' Your nose sticks out and heads the way," she said, triumphantly.

But Tom wasn't so sure.

"We'll end up walking in a circle," he said. "It's well known that people who are lost end up doing that."

"But why?" said Zoe. "I'm sure that we could do it easily."

"Nobody knows for certain why it happens," replied Tom. "It may be that one leg is slightly longer than the other. This means that one stride is a bit longer than the other and you drift away from walking in a straight line. You end up walking in a huge circle and eventually you end up where you started."

"That's stupid," argued Zoe. "I don't go round walking in circles and nor do you."

"That's because you can see where you are going. You make small changes in your direction as you go along because you can see where you are aiming."

"We need something to aim for then," said Zoe. "What about that bright star in front of us. I know that it's North because the door is right behind me. Right, let's go."

"It won't work, Zak," said Tom, before she could set off into the darkness.

"Why not, Tom? You're just looking for things to spoil my idea," said Zoe.

"I'm not, Zak. I like your ideas. In fact, I think you've given me an idea that will work."

Tom looked up into the sky trying to pick out recognisable patterns from the hundreds of stars.

"We can't follow your bright star," he said, eventually, "because it will move across the sky. It might point North now but in half an hour's time it won't."

"The stars don't move, Tom."

"Yes, they do. At least, they appear to move because the Earth is spinning around on its axis. The stars move in the same way that the Sun moves during the day," said Tom.

"Oh, I see," said Zoe, "so if we follow that bright star we'll get lost."

"Yes," said Tom. "But there is one star that always points North. It's called the Pole Star."

"Which one is it? There are thousands to choose from and they all look the same."

"It's easy," said Tom. "Do you see The Great Bear?"

Zoe looked into the sky.

"Do you mean Grizzly?" she said, trying to make a joke of Tom's suggestion.

"It doesn't look much like a bear," said Tom. "It's those seven stars up there."

He held Zoe's head in his hands and tried to point her eyes in the right direction.

"Some people call it The Plough because it looks like an old-fashioned plough. Others say it looks like a soup spoon or ladle," he said. "Look for seven bright stars and see if you can make out the pattern."

"I see what you mean," said Zoe. "They do look like a soup ladle."

She pointed to each star in turn and Tom could tell that she was looking at the right ones because he could see the blackness of her arm as she pointed.

"So which one is the Pole Star?" she asked.

"So which one is the Pole Star?" she asked

"None of them," replied Tom. "But the last two stars in the constellation can guide us to it."

"What's a constellation, Tom?"

"It's a group of stars that make a pattern. There are lots of constellations and The Great Bear is just one of them. It's one of the easiest to spot as well."

"OK," said Zoe. "How do we use the last two stars?"

"Follow a straight line in the sky going through those two stars. The next star that you come to is the Pole Star."

"It's not very bright," said Zoe. "Have I got the right one? You'd think that such an important star would be really bright."

"No, it's not very bright," said Tom. "It just happens to be the only star that doesn't appear to move in the sky, and it always points North. The best way to find it is to find The Great Bear first and then use The Pointers to get to the Pole Star."

As he said this, the eerie sound of crying wolves came through the still, black air. The sound seemed to come from all directions and also seemed much closer than before.

"I think we'd better hurry up," said Tom. "The Crusade doesn't like it if we hang around for too long. It's like having a time limit for each section of the game. Come on, follow that star."

Tom put Grizz onto his lead and they set out into the darkness leaving the stone wall directly behind them. Apart from the wolves howling in the distance, it was strangely quiet around them. Tom could hear the panting breaths of Grizz as he strained on the lead ahead of him. It was odd that he noticed the crunching of their footsteps, but he thought it was probably because he was using his ears more than his eyes for once. Zoe was hanging tightly to his arm as they trudged into the darkness. They both walked in silence, staring upwards at the Pole Star, letting it guide them as it had guided so many navigators in the past.

It seemed like they had been walking for ages but Tom guessed it had only been about twenty minutes, when a breeze suddenly sprang up, blowing into their faces. Dust got into their eyes and Tom found it difficult to look upwards at their guiding star. He found that he had to peer through the fingers of his hand so that he did not lost sight of it. Worse was to come a few minutes later. Cloud rolled in from nowhere and all the stars went out, hidden from view. It was so dark that they could see nothing at all. Grizz continued to pull on his lead.

"Stop Grizz!" said Tom, sharply. "We'll get lost. I've lost my bearings already."

He was beginning to panic. It was pitch black and he had no idea where he was or what direction to go in. Grizz continued to strain on his lead, but Tom yanked him back strongly.

The wind strength increased and they turned around to stop sand from getting into their eyes. From all about them they could hear animals, wolves, scampering around. The wolves were no longer making any noise. They had found their prey and were getting ready for the kill. Tom knew that they had but a few minutes to get out of this mess.

Grizz started to bark and he was tugging at Zoe's trouser leg.

"Grizz knows the way," said Zoe, "and I think I know why. The sign on the door said 'Head North. Follow you nose.' Grizz is following his nose because there must be a scent trail."

"You could be right, Zak," shouted Tom above the roar of the wind. "Dogs have an excellent sense of smell. I read somewhere that it is about a million times better than ours. The clue on the door seems to make sense as too. Well done, sis."

"The only other thing we can do is wait for the clouds to disappear," said Zoe, "and it could be ages before that happens."

"Or it might never happen," said Tom. "If you can't solve the problems in the Crusade then you don't survive. I think we'll take a chance on Grizz before we end up as wolf breakfast."

Tom allowed the lead to go slack and let Grizz drag them along. The little dog ran, without hesitation, into the darkness and straight into the wind. Tom covered his eyes and let Grizz pull him and Zoe forwards. They left the wolves behind them and before long the only sign of them was the occasional howl from somewhere behind in the distance.

Tom peered through his fingers and saw a small light glowing directly ahead of him. The wind began to drop so that he could remove his hand completely and the clouds disappeared to reveal the stars once again. The light was shining from the window of a small hut and directly above the hut Tom saw the Pole Star. They had come the right way.

"I'm sure I can smell pine trees," said Zoe as they neared the hut.

"Yes, so can I," said Tom.

The hut had a door and Tom knocked politely. There was no answer so he pushed the door open and they stepped inside. As they did so the lamp inside the hut went out and it was suddenly as bright as day outside. They looked back over the land that they had just walked across. It was perfectly flat but they could see their footprints coming from the south in a perfect straight line. Beside the trail were a series of miniature fir trees placed every fifty metres and forming an avenue.

"It must have been the fir trees that Grizz could smell all along," said Tom. "Well done, boy."

He scratched the dog under the chin and Grizz jumped up enthusiastically.

"It would have been so easy in the light," said Zoe. "But well done, Grizz," she added, patting him on the head.

They turned around and studied the inside of the hut. It was three metres square with a flat roof and four walls. One of the walls contained the door and the window through which they had seen the light. The two adjoining walls were completely bare and the opposite wall held the familiar computer terminal and monitor. In the middle of the floor was a round hole about a metre in diameter. Tom peered over the edge and saw that the hole went down a long way. He guessed this because in the distance he could just make out a spot of light. If the hole was the same width all the way down then he judged that it must be very, very deep.

"Look, Tom!" shouted Zoe. The sudden noise almost frightened Tom into losing his balance and find out just how deep the hole really was. But he managed to steady himself and he went over to Zoe who was inspecting the computer screen.

"There's another letter," she said.

Sure enough, displayed on the screen were the letters M, E, H and I.

"It's another clue to the password," said Tom. "Not that it helps very much."

He thought for a few minutes and then came up with a possible password.

"HEMISPHERE,' he typed.

"What's that?" said Zoe.

"It's half of a sphere or ball shape," said Tom.

"And what's that got to do with anything?" said Zoe.

"Well, the starry sky looks like the inside of a black-coated ball with dots of light stuck to it," said Tom.

"No chance," said Zoe as Tom pressed the RETURN key.

'ILLEGAL PASSWORD . . . M E H I.'

"Told you," said Zoe.

Tom hadn't held out much hope either. Nor was he too disappointed with their failure to guess the word. He was sure that the only way out was to solve the Crusade first.

Chapter Five - Slipping Up

Zoe got down on her knees and looked down the hole in the middle of the floor of the hut. She saw that the hole went a very long way down so decided it would be more sensible to lie down on her stomach and peer over the edge.

"I suppose we've got to go down there now," she said, timidly.

"You're probably right there," said Tom. "But first I'll take a quick look outside to check for any more clues."

He walked out of the door but as soon as he did so the lights went out and the lamp inside the hut switched on. He looked up into the sky and saw hundreds of stars twinkling brightly. Undeterred, he made his way around the hut, feeling the wall with his hand as he went around. When he was at the back of the hut he could barely see his way. The darkness was like black velvet and it seemed to suffocate him. It was quite scary so he went back to the door as quickly as he dared. He had visions of falling down a hole similar to the one inside the hut so he tested each step gingerly before he put his weight forward. When he got back inside the outside lit up again to reveal the avenue of trees down which they had walked from the spider's lair. He stepped back out and all was dark again. It was very odd and also quite amusing. He stepped backwards and forwards across the doorway. The sky went bright, dark, bright, dark and every time it switched in an instant.

"What's going on, Tom?" asked Zoe, who had joined him at the doorway to try the trick for herself.

"I think you were right about having to go down the hole," said Tom, by way of reply. "I don't see what else we can do. The Crusade is controlling the light and it seems to be saying that we've got to stay in the hut. The only way out of the hut is down the hole."

"Unless we can get the password out of the computer," added Zoe.

"Well we can't get it, can we?" said Tom, a little abruptly.

"Not yet," said Zoe, crawling over to the edge of the hole and peering down it. "It's a long way down and there's no rope or ladder or anything to hang on to."

"I know," said Tom, "but I've been thinking about that. I think we can use friction to help us down."

"What's friction?" said Zoe, looking round in case it was some mysterious object that Tom was referring to.

"Friction is everywhere," explained Tom. "It stops things from moving a lot of the time. You know when you're on your bike riding along a flat road?"

"Yes," said Zoe, a quizzical look on her face.

Tom could see that she was wondering what friction had to do with her bicycle, so he continued with his explanation.

"What happens if you stop pedalling?"

"I slow down and stop. That's obvious," Zoe replied.

"Well, it's friction that slows you down and stops you," said Tom. "There's friction stopping the spinning of the wheels and there's more friction from the air pushing against you."

"So friction is a bit of a nuisance," said Zoe. "If there was no friction I wouldn't slow down."

"That's true," agreed Tom, "but if there was no friction you wouldn't be able to get moving at all because the wheels wouldn't be able to grip the ground. The wheels would slip round and round as you pedalled and you wouldn't go anywhere."

"Oh, I see," said Zoe. "But I don't see how friction can help us to get down that hole."

"Watch," said Tom, confidently. He sat at the side of the hole and dangled his legs over the edge. Then he placed his feet against the inside of the hole, bent his knees and lowered himself in completely with his back pushed up against the whole wall.

"There you are," he said, "Friction is holding me up. All I have to do is lower myself down bit by bit."

"Friction is holding me up"

With these words he started his descent. He may have sounded confident but, in truth, his legs were shaking and he had a sick feeling in his stomach. However, he thought it would be better for Zoe if he made out that he was not worried.

"As long as I push hard with my legs, friction between my back and the hole and my feet and the hole will stop me slipping down. Look, no hands," he said with a cheerful grin on his face and waving his arms up towards his sister.

Zoe stared down at him, swallowed loudly but said nothing.

"Pass Grizz down and then you follow me down," said Tom. Zoe did as she had been asked and rested the little dog onto Tom's lap. When Tom had moved down a bit she lowered herself over the edge. Tom saw her body quivering above him and hoped that Zoe's legs were strong enough to hold her up. If she slid on top of him he wasn't sure he could get the grip needed to hold them both.

"This is easy," said Zoe. "Thank goodness for friction"

For the moment, Tom breathed a sigh of relief.

They seemed to be going down for ages. It was certainly a very deep hole and when Tom looked upwards through Zoe's legs, all he could see was a tiny speck of light. There was no way that they would have the strength to clamber back up there if he had made a mistake. Perhaps he should have looked around for a bit longer before beginning their descent. His worries were interrupted when he looked down.

"Stop!" he yelled, rather too loudly for somebody who was supposed to be in control of the situation.

The shaft had come to an end about three metres above ground level. Bolted across the end of the hole was a metal bar and this had stopped Tom from going down any further.

"Hold on to Grizz for a moment," he said, and he passed the bewildered-looking dog up to Zoe. He then squeezed past the bar, gripped it tightly and lowered himself down like he was hanging onto a trapeze. This way he had only about a metre to drop. He let go of the bar and landed on the ground, slipping over immediately. He seemed to have landed on an enormous ice rink. With great difficulty because he kept slipping, he scrambled to his feet and stretched up his arms towards Zoe.

"Drop Grizz to me," he said.

Zoe followed the instruction without hesitation and the dog flew through the air and into Tom's arms. The impact made him step to one side, he slipped and they both fell to the ground.

"Ouch!" cried Tom.

He put Grizz onto the ground whereupon the little dog's legs splayed outwards so that he rested on his stomach. Grizz tried to get up but couldn't move. He began to whimper.

By now, Zoe was hanging from the bar and ready to drop.

Zoe was now hanging from the bar

"Let go!" said Tom. "I bet you can't stay on your feet."

She dropped and did the splits as she landed and so all three of them were now lying on the icy material. They were a mass of arms and legs, slipping and sliding all over the place.

"This is the most slippery ice-rink ever," said Tom laughing as he watched the others trying to get up. "We could definitely do with some friction here."

By this stage Grizz had given up and was a very sorry sight. He lay there with his head resting on the ground, panting heavily.

After some minutes, by which time Tom no longer found the situation funny, they managed to get upright.

"It's not ice, is it?" said Zoe. "It's not cold or wet."

She was quite right. It wasn't ice. Although it looked like ice, it was far more slippery. It was some kind of material that was perfectly smooth and harder than rock.

"Look, there's the computer screen," said Tom, pointing to his right. "That's where we need to get to."

About five metres away they could see a keyboard and monitor next to a set of steps leading up to some sort of tunnel.

"But how do we get to it?" asked Zoe. "I can't move. Every time I try to take a step I just slip. I don't get anywhere at all."

She tried to demonstrate as she spoke. It looked quite funny but Tom realised that it was quite serious. With nothing to grip on to it was impossible to make any progress at all.

"It's odd, isn't it?" said Tom. "You need a force to get yourself moving and the force normally comes from friction. If there's no friction you cannot move."

"So how can we get away from here?" asked Zoe.

Tom thought hard. After a moment he said, "I could push you."

"But you'll still be here," said Zoe.

"I know," said Tom, "but maybe there's something by the steps over there that you could throw to me. A life-line or something. Can you see anything?"

They looked over towards the computer. It was fitted into a cliff face and there were steps leading into a tunnel.

"There must be something in the tunnel. There's bound to be, otherwise the task is impossible," said Tom, rather puzzled and not at all convinced.

He told Zoe to sit down and he placed Grizz into her lap.

"I'm going to push you very gently. When you get to the computer grab the railing by the steps," he said.

And without considering the problem any further, he pushed.

Zoe slid slowly, but surely, towards the computer terminal. When she got there she grabbed the railing and placed Grizz safely on the first step.

"Stay!" she ordered, but Grizz didn't look like he wanted to go off adventuring anywhere.

Meanwhile, Tom was realising that he had made a terrible mistake. Sure enough, he had pushed Zoe and Grizz to safety but, in doing so, an equal and opposite force had pushed him the other way. It was like firing a cannonball. The cannonball goes one way and the cannon backfires the other way. Admittedly, he was moving very, very slowly, but he was definitely moving away from Zoe and Grizz. Worst of all, there was nothing he could do about it.

"Zoe, quick, help!" he shouted. "Throw me a line. I'm moving away from you."

Zoe turned round to see Tom moving very slowly away from her. She acted quickly, running up the steps, looking for something to throw. There was nothing to hand.

"There's nothing!" she shouted.

"Look at the computer screen," shouted Tom, starting to panic. "Maybe we can guess the password. It's our only chance."

Tom knew that there was no way that he could stop moving. If there was no friction he couldn't start himself moving. He had already found that out. But if he was already moving there was no way that he could stop. He would keep on going very slowly in a straight line forever or until he hit something. He looked where he was heading. The smooth, slippery surface seemed to go on as far as he could see and that was where he was going.

Zoe was now about ten metres away and she was staring at the computer screen.

"There's another letter," she said. "It's M E H I C. What's the word, Tom?"

Tom thought hard. It was his only chance. But for every second that he thought he was moving a little bit further away from safety.

"Calm down," he whispered aloud to himself and he drew in a few deep breaths. "M E H I C. Think of a scientific word with those letters."

Tom was sure that the password was something to do with science.

His brain was working overtime and then a word came to him.

"Chemistry!" he shouted. "It's CHEMISTRY."

Zoe turned to the keyboard. 'KEMISTREE' she wrote and then pressed the RETURN key.

'ILLEGAL PASSWORD . . . M E H I C'

"It doesn't work, Tom," she called. "But I didn't use all of the letters anyway. I didn't need the H or the C."

Tom realised immediately that chemistry was not an easy word for Zoe to spell. By now he was fifty metres away and still moving of course, sitting on the slippery ice-like substance.

"C, H, E, M, I, S, T, R, Y," he shouted. He wondered if Zoe could hear him or type the letters fast enough. He repeated the spelling, leaving a gap between each letter as he shouted it.

In reply, he heard a choked voice shouting back, "It doesn't work, Tom."

Tom swallowed hard himself. It was a strange feeling watching his sister disappearing into the distance. Perhaps he would never see her again. Before he could dwell too long on that thought, he realised that he was no longer alone.

His attention was distracted by an octagonal metal object about one metre across and half a metre high that was sliding along beside him. Every so often it made a hissing sound from one of eight small pipes that were placed in each of its eight sides. It was blowing air out of these to control its movement, just like a rocket whizzing around in outer space. It sucked in air from a small tube sticking out of the top of the machine and then blasted it out of the side pipes. Squatting at the controls was an octopus of some kind. Its eight tentacles were resting against the eight buttons that controlled the movement of the machine. To move a bit to the left it pushed the left button. "Pssst." Air was sucked in through the top pipe and blown out of the right pipe. The machine moved a fraction to the left. The word 'OCTOPUSH' was written along one side panel.

Squatting at the controls was an octopus of some kind.

"Ingenious," thought Tom, "it's the only way to get around in a frictionless world. Jet propulsion."

"You failed," said the creature, unexpectedly, "and soon it will be the end of the game for you."

"What do you mean?" asked Tom, politely. The beginnings of an idea were forming in his head and he wanted to stay on good terms with the Octopush.

"You cannot stop, you know. You're moving at ten point eight centimetres every second and in precisely seventeen minutes you will come to the end of the slippery rink.

"What's at the end?" asked Tom.

"A sheer drop." replied the Octopush, in a matter-of-fact sort of way, "into a sea of acid. End of game."

It chuckled, then made hasty adjustments to its direction. "Pssst, pssst," it sounded.

"You should have brought the rope with you, " it continued. "Then you could have pushed your friend to the steps. The rope unravels and then your friend hangs on to the railing and pulls you in like a fish on a line. As long as you don't pull too hard the rope will not break. But you failed!" it squealed.

The eight-legged creature started to laugh.

"It's always the same," it said. "Most people get this part wrong."

The Octopush skilfully manoeuvred the machine around Tom as it roared with laughter. Air hissed out of the jet pipes controlling the movement perfectly.

"You are now moving at ten point seven centimetres every second and you die in sixteen minutes," it said.

"I'm slowing down slightly," said Tom.

"A little," said the Octopush. "It's because of air resistance but I'm very sorry to say that it won't slow you down enough."

Tom took off his jacket and held it out like a big, red sail, trying to increase his air resistance.

"Very good," said the Octopush. "That will slow you down a little more so . . . let me work this out. You will die in sixteen minutes and ten seconds precisely."

"Pssst," the Octopush made a small adjustment to its own speed so that it kept alongside Tom. Tom put his hand in his jacket pocket. He felt the familiar objects. There was the bubble gum, the magnifying glass, the compass, the balloon and some scraps of paper. He unwrapped the gum, popped it into his mouth and thought hard.

"Pssst," went the Octopush and Tom had an idea. He put his jacket back on and took the balloon out of his pocket. He blew it up as far as he dared without risk of bursting it, then turned it around and let the air out. If the Octopush could do it then so could he. He would jet-propel himself back to Zoe and Grizz.

The Octopush was almost hysterical with laughter. In between giggles it reported, "You are now travelling at ten point five centimetres every second. You need to blow the balloon up over one hundred times in the next twelve minutes to stop yourself moving. You'll never do it but nice try."

Perhaps the Octopush was right but Tom wasn't going to give up just yet. He blew the balloon up again and released the air, deliberately letting it make a raspberry noise. The Octopush was losing control of itself. The entertainment was never this good. Most victims screamed and pleaded for help. Some offered money. But this one was blowing up party balloons, letting the air out and playing a tune at the same time. It was hilarious.

Tom noticed that his antics were causing the Octopush to lose a bit of control of the machine and another idea came into his head. He continued to blow up the balloon and distracting the Octopush.

"Ninety-eight times in ten minutes and forty seconds," it said, enjoying the show.

It was then that Tom and the Octopush collided. Tom took the gum out of his mouth and rammed it down the tube sucking in the air. He then pushed as hard as he could against the machine with his legs. It worked.

In an instant he had stopped and then started travelling back the way he had come. The Octopush was now hurtling towards the edge of the slippery rink. Its air intake tube was blocked so it could no longer control its movement. It would carry on in a straight line at a constant speed until it slipped over the precipice at the edge of the rink.

"Enjoy the acid," shouted Tom. "You'll be there in approximately three minutes." He judged this by looking at the speed with which the Octopush was moving away from him.

The laughter of the Octopush had turned to a high-pitched scream of terror.

Tom was making good progress back the way he had come. There was no friction force to stop him getting back across the rink, apart from a little air resistance. And that would hardly slow him down at all. He felt very pleased with himself now so he sat back to enjoy the ride.

He had made a stupid mistake earlier on but had managed to solve his problems like a true scientist. Actually, he quite enjoyed being in a frictionless world when he could lie back and get from one place to another with no effort at all.

He looked up to see his sister in the distance. She was sitting on the steps holding her head in her hands.

"Zak!" he shouted when he was close enough to be heard.

He saw her look up and then she jumped to her feet and started waving. He waved back. He couldn't remember being so happy to see her in his whole life. He got to the steps and grabbed the rail. Zoe had her hands around his neck and was hugging him and Grizz was jumping up and licking his face.

"Get off, Zak," he said. "Let me see the screen." He tried to appear cool and calm but inside he was shaking with relief.

On the screen he saw the flashing password message.

'CHEMISTRY,' he typed, just in case Zoe had spelt it incorrectly when he had shouted the word at her earlier.

'*********' appeared on the screen. He pressed RETURN.

'ILLEGAL PASSWORD . . . M E H I C ' flashed the reply.

"Come on, let's go up there," he said turning to walk up the steps. He was no longer too concerned about the password. It was too difficult to guess. They needed more clues.

So they climbed the steps and walked into the tunnel in the cliff face.

Chapter Six - Gravity Pulls O.K

The tunnel was quite long but it was not dark even when they had walked well into it. It was like an art gallery. On the walls were pictures of famous scientists and each one had a name plate and a small piece of information. Tom recognised one or two of the names and he stood and looked in wonder at some of the famous people. The paintings were so good that it was like seeing the actual person in the flesh. They were also holograms so that the image looked three-dimensional. He came to a life-size portrait of Galileo, the Italian astronomer, and he pointed it out to Zoe.

Tom recognised one or two of the names and stood in wonder.

"This is Galileo," he said, "one of the most famous scientists of the sixteenth century."

"I can read, Tom," said Zoe, looking at the information label.

"Yes, but I was going to tell you a bit more about him," said Tom, a little hurt.

"Sorry, Professor Kennedy. Go on."

"He had lots of ideas," said Tom, "but there was one idea that got him into a lot of trouble. He said that the Earth went round the Sun and not the Sun around the Earth. Nobody believed him at the time and in the end he had to say that he had changed his mind."

"Why?" asked Zoe.

"Because otherwise he would have been killed. Everyone else thought that he was mad."

"That's silly," said Zoe. "Even I know that the Earth goes around the Sun. Does that make me a great scientist too?"

"I'm afraid not," said Tom. "To be a great scientist you have to think up something that nobody else has thought up, and then you have to show that your idea works."

They continued their walk into the tunnel. They came to a picture of Isaac Newton.

"He's the one who sat under an apple tree and an apple fell on his head," said Zoe, laughing at the rather stern-looking man in front of her.

"We don't know if that really happened," said Tom, impressed by his sister's knowledge. "But it's quite likely that he saw an apple fall to the ground and that's why he came up with his idea of gravity."

"What exactly is gravity?" said Zoe.

"Gravity is what pulls you down to the ground, otherwise you would float away," said Tom, unsure how to explain it any better, but pleased that Zoe was taking such an interest.

"Look, there's Einstein," he said, changing the subject before Zoe could ask him a more complicated question. There were so many questions that he had no answers for. And the strange thing was that the more he found out about science, the more unanswered questions he had.

They stood for a moment looking up at the hologram of Albert Einstein.

"Now Einstein was a brilliant scientist," said Tom. "He thought of some ideas that he couldn't even test out for himself because he didn't have equipment that was accurate enough. It was only years later that other scientists found out that he was correct."

"What ideas were they?" said Zoe.

"I'm not sure exactly, but I know they were about space and time," said Tom, frustrated by his own lack of knowledge. "That's the next thing that I want to find out about if I can. As long as we can get out of here," he added.

There were hundreds of pictures of other scientists. There was one of Marie Curie, one of Edward Jenner, one of Ernest Rutherford and many, many others. Tom had to admit that he hadn't heard of most of them, and even if he had heard of the name, he didn't know why they were famous. Reading the information tag didn't help much either. He simply didn't understand what the writing was going on about.

Towards the end of the tunnel there were a series of blank holograms.

"Those must be for the next great scientists," said Tom.

"Well, there must be lots of ideas still to be thought of because the blank frames go on for ages," said Zoe.

Tom thought that she must be right. Even though it seemed as though everything had already been invented or discovered, it couldn't be true. He wasn't the only one with questions that had no answers. Tom stared at one of the blank frames and imagined himself inside it. 'Tom Kennedy - famous' . . ."

"Tom, quick, over here."

Tom came out of his daydream with a start. He saw that Zoe was at the end of the tunnel and he ran over to meet her. The tunnel opened into a huge cavern and Zoe was standing on a ledge about eighty metres above the ground. It was a sheer vertical drop to the rocks below. It made them dizzy to look down so they lay on their stomachs and peered over the edge. Grizz shrank back into the tunnel with his tail between his legs. The cavern was lit by an enormous chandelier which was hanging from a single chain fastened to the roof. It was swinging in the breeze that was blowing up the tunnel behind them, casting strange eerie shadows on the walls.

Suddenly, as they looked downwards, there was a sound of rushing air and a huge, inflatable bag expanded at the base of the cavern. It was like the crash mat for a pole-vaulter but much bigger. After a couple of seconds it deflated just as quickly and all but disappeared from view. It went quiet again except for the rhythmic creaking of the swaying chandelier.

"I've got a feeling that you're not going to like my next suggestion," said Tom, interrupting the hypnotic effect of the swinging light. "We've got to jump down there." He nodded towards the bottom of the cavern.

"Look, there's the computer screen and keyboard at the base of the wall," he said, pointing across the cavern. It was a long way away but, sure enough, there was a monitor glowing in the distance.

"No way," said Zoe. "There's no way I'm jumping down there and that's final."

"But it's the only way out," argued Tom. "We can't go back the way we came and the next level of the game is down there. We've got to jump, Zak."

"We'll die. Nobody can survive a jump this high," said Zoe. "There must be another route or else we made a mistake coming up here in the first place."

"It'll be alright," said Tom. "Just close your eyes and leap out. It's quite safe because we'll land on the air bag. That's what it's there for."

"But it's not there anymore, is it Tom? It was only there for a couple of seconds and now it's gone," said Zoe, quite correctly.

"Well, when it comes back, we'll jump," said Tom, extremely calmly considering the suggestion that he was making. But he was getting used to dangerous situations now and he knew that the tasks were not going to get any easier as they went through the levels of the Crusade.

At that point, there was another rush of air as the giant bag inflated again. Two seconds later, the air gushed out and the bag was flat.

"There is a slight problem, of course," said Tom, having made careful observations of the air bag. "I think it'll take quite a long time for us to fall down that far. If we wait for the bag to blow up and then we jump . . ."

". . . the bag will be gone by the time we land," said Zoe, finishing Tom's sentence.

"It's a bit like the cliff divers of Mexico," said Tom. "They have to time their dive so that they hit the water when a wave raises the level of the sea. If they get it wrong the water is too shallow and they get badly injured or killed."

"Let's test it out then," said Zoe, getting to her feet.

For a moment, Tom thought that Zoe was going to make a jump for it but, to his relief, she turned away from the ledge and went back into the tunnel. Shortly afterwards she came back with a large rock in her arms. It was quite a large, smooth rock and she was struggling to lift it. Tom could see what she was going to do and he helped her to carry it to the edge of the drop. They stood and waited patiently for the air bag to inflate. It seemed like an age but only because the rock was heavy and it was making their arms tired.

'Whoosh.'

They heard the air rushing into the bag.

"Now!" said Tom, and they threw the rock over the edge. They got down onto their hands and knees and watched the bag inflate, then they saw it deflate, and then they saw the rock landing on the solid ground. It smashed into hundreds of pieces with a crash.

"Just as I thought," said Zoe. "What now?"

"It's obvious," said Tom. "We jump before the bag inflates. When we're part of the way down, the bag will fill with air, we land safely and walk away without a scratch."

"But how are we going to know when to jump?" said Zoe. "You know what happened last time when you didn't think things through properly. How are we going to know?"

'Whoosh!' The bag inflated again. 'Wheesh!' And then it was gone.

Tom's mind had already been working on the problem.

"I reckon that the bag blows up at certain times," he said. There are an exact number of seconds between each inflation. All we have to do is see how long it takes and then we'll know when it's safe to jump."

As he said this, he lifted his cuff to look at his digital watch. His heart sank when he realised that he did not have his watch on. His mother had taken it into town to get a new battery for it. And that was why he had to look after Zoe, and that was why he was caught up in this mess. He held his head in his hands and groaned.

"What's wrong now, Tom?" said Zoe.

"Have you got a watch on by any chance?" asked Tom, knowing full well that Zoe didn't have a watch.

"No," she said. "You know I haven't got one."

"We'll have to count the seconds then," said Tom, but he didn't have much faith in his ability to get it right. He'd tried it before. He'd closed his eyes, set the stopwatch going and tried to guess when a minute was up. He never got it right. Sometimes he stopped the watch too early, sometimes too late. Sometimes he got quite close, but what use was that if he couldn't be sure when he was going to get it right. But Tom couldn't think of any other way that they could tell when to jump over the cliff.

'Whoosh!' The bag inflated again and Tom started counting the seconds out loud.

"One and two and three and four and . . ," he chanted, trying to strike up a one second rhythm. When he got to thirty-three the bag inflated again.

"One and two and three and four and . . ," he counted, trying the test again. At the count of thirty-nine the bag blew up once more. Tom started counting again.

He tried counting the time gap on five more occasions and he got a different number each go. The lowest number was twenty-eight and the highest number was thirty-nine.

Tom sighed. If only he had his watch. And Zoe wasn't exactly being helpful. He looked over at her and saw that she was staring into the cavern. The chandelier was casting strange shadows as it swung back and forth in the chamber and Zoe seemed almost hypnotised by the effect. Tom nudged her.

"You could at least help," he said. "Why don't you try counting as well?"

"I wondered what you were doing," said Zoe. "You never explained and I thought it best to keep quiet while the great scientist was hard at work."

"Well, I've been trying to get us out of this mess, if you must know," said Tom, rather crossly. "I'm counting the seconds between when the bag blows itself up each time. The trouble is, I'm not a very accurate timepiece."

"Shall I try," said Zoe.

"No, there's no point. I suppose I could take the average number of seconds that I counted but it's far too risky. The bag is only there for a couple of seconds and there's a good chance that we'll miss it."

"I've got an idea, Tom," said Zoe.

"Oh, so there is life in there," said Tom. "You looked like you were in a trance just now, staring at the chandelier."

"While you were doing your counting I noticed that the bag always blows up when the chandelier is furthest away. Look, it swings away and back, away and back, and on the eleventh time the bag appears."

At that moment, the chandelier moved away from them and there was a whooshing sound as the air bag inflated below them.

"Let's count the swings," said Zoe. "One!" she began, as the chandelier moved away again. They continued the count together.

"Two," they said.

"Two and a half," said Tom, as the huge light moved up to them.

"Three," they said together.

When they shouted, "Eleven!" the bag inflated below just as Zoe had said.

"Well done, Zak," said Tom, excited by his sister's discovery.

At that moment there was a bigger gust of wind from along the tunnel. It was so strong that it almost blew them over, and the chandelier started swinging a little bit more.

"Oh no!" exclaimed Zoe. "That will mess everything up. It's swinging much more now." She shrugged her shoulders with obvious disappointment.

"No, it won't, Zak. Listen," said Tom. "You know that picture of Galileo that we saw in the tunnel?"

"Yes, I think so," said Zoe. "At least I remember the name."

"Well, he did exactly the same thing as you. One day he watched a chandelier swing back and forth. But he also noticed that the size of the swing made no difference to the time of the swing. If the swing was a bit bigger then the chandelier seemed to swing a bit faster. You know what it's like when you get on a swing yourself. The bigger the swing, the faster you go. But the time for the swing does not change. The only way in which you can change the swing time is to make the chain longer."

"Let's test it out again, then," said Zoe.

The next time that the bag swelled up with air, they began to count again. After eleven swings the safety bag inflated.

"So, if we jump after ten and a half swings, we'll land just as the air bag blows up," said Tom. "And here is the proof."

Tom had picked up a stone while he was making his prediction and he had been keeping a careful count of the swings.

When the chandelier reached the near side on one of its swings, he dropped the stone. About halfway through the descent there was a whoosh of air and the bag ballooned out. The stone landed gently in the middle of the bag.

"Brilliant!" said Tom, jubilantly.

"We'll jump in two goes time, and we'll count the swings together out loud so we don't make a mistake," he said, hurriedly.

"You're doing it again," said Zoe.

"What?" said Tom.

"You're rushing things. We're different sizes, aren't we? That means that you'll go down faster than me and I'll go down faster than Grizz. You only tested it out with a small stone."

Tom was half expecting this from Zoe. He wasn't going to mention anything about it unless she had asked, but he had to admit that she was beginning to think like a real scientist. Anyway, he had an answer ready. He picked up a large rock and placed a small stone beside it.

"It doesn't matter how big you are," said Tom. "Everything falls to the Earth at the same rate. Look at these stones." He pointed to the rocks that he had collected.

"This big rock needs a big force to get it moving and this small stone needs a small force to get it moving. Agreed?" he said.

To demonstrate he pushed the large rock quite hard with the palm of one hand, and he moved the small stone with the forefinger of his other hand.

"Yes," said Zoe. "And . . . ?"

"And gravity pulls on the big rock with a big force and gravity pulls on the small stone with a small force. The result is that both objects get moved at the same rate."

He pushed both rocks over the ledge. They looked over and saw them land at exactly the same moment.

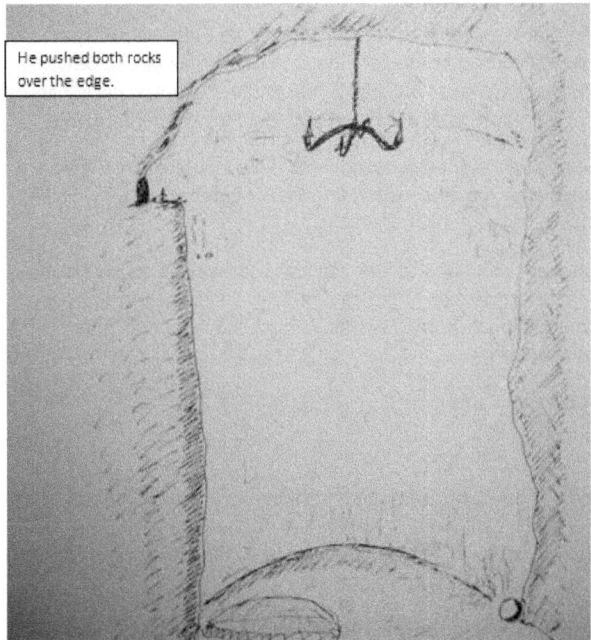

He pushed both rocks over the edge.

"So we jump at the same time!" said Zoe, excitedly.

"That's right," said Tom, who was feeling more confident with the plan himself now that he had explained it. He often found that when he had to explain an idea to somebody else it helped with his own understanding. He was now fairly certain that his idea would work, but he had one nagging doubt.

"The only problem may be air resistance. When we jump we must both roll up into a tight ball and fall backwards. That way we'll have roughly the same drag force but at this height I don't think that it'll make too much difference."

They discussed what they were going to do for quite a long time. Tom thought it would be best if Zoe held on to Grizz and he would shout "Jump!" at exactly the right moment. He was very serious when he told Zoe how important it was to jump when he said. She must not hesitate or have second thoughts.

They checked the swing of the chandelier for eight more inflations of the air bag. Then they decided that it was time to do the jump.

They waited for the next inflation and then counted the swings together.

"Half . . . one . . . one and a half . . . two . . . ," they chanted.

Tom could feel his stomach churning and Zoe had gone very pale. They carried on with the counting.

"Nine . . . nine and a half . . . ten . . . ten and a half. Jump!"

They leapt out into thin air and gravity pulled them down towards the Earth below. Tom looked down and saw solid ground coming up to meet them. He looked across at Zoe clinging to Grizz, her teeth clenched and eyes bulging. A thought flashed through his mind that they were falling at exactly the same rate. It didn't matter that they different weights.

"Whoosh!" They heard the air bag inflating and a second later they landed upon it. It was as soft as a cushion. Almost immediately the air escaped from out of the bag and they hit the bag with the slightest bump.

Tom tried to stand but his legs were so wobbly that he couldn't. Zoe was obviously feeling the same so they crawled to the opposite side of the cavern. Only Grizz seemed to have survived the experience without too much worry. He was jumping around, barking and yapping as though he wanted another go.

"How did you like free fall, Zak?" said Tom.

"I didn't," she replied.

The computer screen flickered beside them. Tom got to his feet and looked at the picture. He saw his bedroom as before and he wondered what time of day it was and whether his mother was worried about them. It felt as if they had been in the adventure for days, but in computer game time it might not have been very long at all.

'PASSWORD,' flashed in the top right corner and there was another letter clue to help them. The sequence now read, 'M E H I C A.'

'MACHINE,' typed Tom, and he pressed RETURN.

'ILLEGAL PASSWORD . . . M E H I C A.'

"How much longer do we have to go?" asked Zoe.

"I don't know," said Tom. "I don't know how long the password is but it can't be much longer. Every time we solve a problem we get another letter clue. So far we've earned six letters and passwords are not usually longer than that."

"Perhaps it's a word we've never heard of," said Zoe.

Tom hoped that she wasn't right.

"There is a way that I can check that," said Tom, thinking hard. "I could try every order of the six letters that we've been given so far. Let's see. There are six letters and they can be put into any order. We worked this sort of thing out in maths the other day. The total number of different words is six times five time four times three times two times one. That's . . . . . loads."

"How many is loads?" said Zoe.

"Seven hundred and twenty," said Tom. "It'll take ages to type them all in, but I suppose I'd better start."

"There's no need," said Zoe.

"Why not?" said Tom.

"Well, you said that we got a letter every time we solved a new problem, right?" said Zoe.

"Right!" agreed Tom.

"Well, Grizz has just found us another problem so the word must have at least one more letter."

"I agree," said Tom, looking over his shoulder to see what Zoe was staring at.

Grizz was squeezing into a small hole in the rock face. When he finally got through to the other side they could see daylight through the hole. There was also another gap higher up. This was obviously the exit tunnel, thought Tom to himself. Perhaps it was even the way out of the Crusade.

Unfortunately, the exit tunnel was blocked by a huge boulder. It didn't quite block the tunnel completely, but there was no way anything bigger than Grizz could squeeze through.

Tom put all his weight against the boulder and pushed. It didn't even move a fraction of a millimetre. Zoe helped and they pushed with all their strength. It moved about a centimetre but it was nestled into a dip in the floor like an egg in an eggcup. As soon as they stopped pushing, the boulder rolled back into the same spot as before.

"We're not strong enough to solve the Crusade," said Zoe, her face purple with the effort of pushing.

"Oh, but we are," said Tom. "You need brains, not brawn to solve it."

With that, he sat down to think.

Chapter Seven - Leverland

"What we need is a lever," said Tom, eventually.

"Have you got one in your pocket?" said Zoe.

"I don't think so," said Tom, pretending to search through the pockets of his jacket. "And I definitely haven't got one big enough to shift that boulder." He laughed out loud.

"Don't laugh, Tom," said Zoe. "I don't know what a lever is."

"Sorry, Zak," said Tom. He really was sorry too because, once again, he had been nasty at his sister's expense.

"Levers help you do a job with less effort," said Tom, in his knowledgeable voice. "You can use a screwdriver to lever the lid off a tin of paint. You certainly wouldn't be able to do the job with just your fingernails."

"So what we need is a giant screwdriver to level the boulder away from the escape tunnel," said Zoe.

"Yes, sort of," said Tom. "But any long and strong piece of metal will do."

"Like this," said Zoe, wiping the dust from the ridge she had been sitting on. She revealed a shiny metal girder a bit like a section of railway line.

"Yes, exactly like that," said Tom, leaping to his feet.

Before long, they had cleared the dirt and sand from the whole length of the metal bar. It was about three metres long altogether and about two centimetres thick. It was also extremely heavy but by lifting one end at a time they could just about move it around.

"This is our lever," said Tom. "And now all we need is something to rest it on."

He looked around the cavern and finally spotted a rock that he thought would do the job.

"This is our pivot," he said, placing the rock next to the huge boulder that was blocking their exit.

"Our what?" asked Zoe, looking closely at the rock.

"Our pivot," said Tom. It's a bit like the hinge of a box lid. This is what the lever will turn upon, just like a lid turning on a hinge when you lift it up."

"I'm not sure that I know what you mean again," said Zoe. "But I'll do what you say for the moment."

"We need to put the lever onto the pivot, but with the end sticking out a little bit. The end part goes under the big boulder," said Tom.

Together, they placed the lever into position. Tom made a few minor adjustments by moving the pivot slightly and placing the end of their lever right under the boulder. When he had finished, the long end of the lever was sticking up about two metres in the air.

"Do you want to move the boulder, Zak?" said Tom. He could tell that his sister was not exactly impressed with what they had done so far.

"How?" she said, hands on hips.

"Reach up and pull on that end," said Tom, pointing to the end of the lever. "Stand on that other rock so that you can reach."

"Reach up and pull on that end," said Tom.

"Why have I got to do all the hard work?" said Zoe.

"You haven't," said Tom. "Just do what I say."

"I'll hang there like a monkey and you just want to have a laugh. But OK, I'll do what the foreman says."

She clambered onto the rock that Tom had indicated, reached up to the end of the metal bar, and pulled downwards with all her weight. The end of the lever went down, the other end went up, and the huge, blocking boulder rolled out of the way.

"Wow!" said Zoe. "I moved it, and it moved so easily."

"Well done, Zak. Now just stay there leaning on the bar. Don't let go whatever you do."

Tom got down on all fours and began scrambling through the gap between the boulder and the wall. He could just about crawl through, looking like a worm squeezing into a worm hole.

"You can let go now," he said when he was safely through to the other side. Grizz was there waiting for him and the little dog started licking his face.

Zoe let go of the lever. It sprang back up into the air and the huge boulder rolled back into place. The metal bar then fell to the ground with a clang.

"Tom!" Zoe screamed. "I'm trapped."

Tom was feeling very guilty on the other side of the boulder. He thought that this would be the end of the Crusade but now realised that this was not the case. There was no computer screen on the other side of the gap so that could only mean that the problem wasn't yet solved. His first problem, though, was to get Zoe through the gap as well.

"Zoe!" he called, through the small opening at the base of the boulder. "You've got to push the lever through. Lift one end and swing around until you can push the lever through the hole."

Tom could hear panting and groaning on the other side of the opening and eventually the end of the lever appeared through the gap. Tom pulled it through and then set up another rock to act as a pivot. Before long, he had levered the huge boulder out of the way so that Zoe could squeeze through. First her head peeped through and the rest of her body followed.

"It looks like the rock-face is giving birth to you," laughed Tom as Zoe popped out.

"It pongs out here," said Zoe, by way of reply.

"Thanks very much," saidTom, jovially, but he had to agree. He had not really noticed it before but there was a definite smell of rotten egg sandwiches. What was more, the smell seemed to be getting worse.

They turned around to discover the source of the choking, sulphur-like fumes. In front of them was a hot, yellow river. It was about one hundred metres wide and it was bubbling and rushing past them. There was a bridge crossing the river that stretched from the opposite bank but it didn't quite make it to their side of the river. It stopped about ten metres away on a small island sticking out of the river. Between them and the island was a plank of wood that was perfectly balanced like a seesaw. The pivot for the seesaw was on the island itself.

Between them and the island was a plank of wood that was perfectly balanced like a seesaw.

"This is easy," said Zoe. "Even I can see what we have to do next. We walk along the seesaw plank to the middle, then we jump down onto the island. From there, we can use the bridge to get across this smelly river."

She moved forwards to step onto the plank.

"Wait!" warned Tom, grabbing her arm and pulling her back. He had noticed that the plank didn't quite reach the bank of the river that they were standing on. There was a gap of about half a metre.

"If you stand on there," he said, "the plank will move down and you'll end up in the sulphur river." He pointed to the boiling liquid two metres below them. "We need to pull the plank onto the bank so that it won't be able to drop when you stand on it."

He leaned over and grabbed the plank with both hands but it wouldn't move. It was secured to the pivot on the island.

"If I run I might make it," said Zoe, hopefully.

"I don't think so," said Tom. "The seesaw is very delicately balanced."

He placed a rock about the size of his foot onto the end of the plank. The plank immediately tipped downwards and the end of it disappeared into the bubbling, yellow liquid. The rock fell off and the seesaw balanced itself again.

"What now?" said Zoe. "How can we stop the seesaw moving down when we stand on it?"

"We use the lever," said Tom, a plan forming in his head. "Now, let me work this out because it's not as simple as it first seems."

"Well, you'd better hurry up," sad Zoe.

"Why?" said Tom, beginning to scratch some drawings into the dust.

"Because the river is rising fast," said Zoe.

Tom looked over the edge of the bank and saw that Zoe was right. The sulphur river was rising and it was obviously getting hotter because it was bubbling much more as well. The stench of the gases was choking and he put his nose inside his jacket to filter some of the fumes. Zoe did the same and even Grizz tried to hide his snout inside his neck fur. Tom went back to his sketched plan.

"It's easy enough to get one of us with Grizz to the island but I can't work out how we can all get there," he said, after a few minutes thought.

"Explain what you've worked out so far," said Zoe, her words muffled as she spoke through her jacket.

"Well," said Tom, "if we put the metal bar under this end of the seesaw we can stop the plank from moving downwards. I could lean on the bar and hold you when you stand on the plank. You could carry Grizz to the island while I take your weight with the lever."

"I see," said Zoe. "So far, so good. But how do you get across?"

"I don't know," said Tom. "That's the hardest part of the problem. One answer is for you to stay here while I go across. I could find the next computer monitor and hopefully I'll be able to solve the password clue."

"No way, Tom!" said Zoe. "You're not leaving me behind again. Besides, there's no time. Look at the sulphur river."

Tom turned to see that the river was almost up to the top of the river bank.

"Anyway," said Zoe, "there has to be an answer to the problem otherwise the Crusade wouldn't be fair."

Tom knew that Zoe was right. There had to be a scientific answer but his mind just wasn't thinking straight. The first trickle of molten sulphur lapped over the river bank. As it cooled, it solidified.

"I know!" said Zoe, suddenly. "Why don't I walk to the other end and then we can balance the plank like a seesaw?"

"Good idea, Zak, but you can't get to that end without it tipping down into the river on the other side of the island," said Tom.

"Come on, Tom, think. You could stop it tipping by standing on this end," said Zoe.

Tom still wasn't convinced.

"The trouble is," he said, "because I'm heavier than you, as soon as I stand on this end it will still tip down. I'll tip into the river, and then you will when I've fallen off."

"But not if I had Grizz with me. Together we must weigh the same as you," said Zoe, excitedly.

"Brilliant, Zak. You're right," said Tom with admiration. "Let's test the idea out using the metal lever as a seesaw over that rock."

"We haven't time, Tom. Look!"

The river was beginning to burst its banks just upstream of them. Zoe picked up Grizz and stepped towards the plank.

"Come on, Tom," she said, impatiently. "We've no time to lose."

Tom placed the metal lever underneath the seesaw plank and then stood on the other end of it. Zoe, with Grizz under her arm, stepped onto the plank and started to walk along it towards the island. The sulphur fumes enveloped her figure and she almost disappeared from view at times. Tom heard her coughing and spluttering ahead of him. When she got to the centre of the plank, right over the pivot, Tom found that he no longer had to push down on the metal bar.

"Zak!" he shouted into the mist.

"Yes, I'm in the middle," came the reply.

"Keep walking the plank," shouted Tom, chuckling to himself because of what he had said. It was like an order to Captain Hook in a pirate film.

Zoe edged along the plank and, as she did so, Tom put one foot onto the plank himself. The further she went, the more weight he had to put on the plank. Eventually, he had to step fully onto the plank with both feet. The plank touched the surface of the river but didn't go under. They were perfectly balanced.

"Stop!" shouted Tom. "Turn around."

Zoe turned to face him. She looked terrified, standing on a small plank of wood just above a raging torrent of boiling sulphur.

"We've got to walk towards the middle now," said Tom. "But walk slowly. If I say 'Stop', you must stop."

Zoe started to walk towards him. As she did so, Tom felt the plank drop slightly. To keep the balance, he took a small step towards the middle. The plank went back to the horizontal position. They carried on like this until they met at the centre of the seesaw. From there, it was easy to jump down onto the island.

"We did it," said Zoe, a huge grin stretching across her face.

"Yes, we did well, didn't we?" said Tom. What he meant to say was that Zoe had done well. After all, she had done most of the solving and the more dangerous part as well.

"Let's get across the bridge. These fumes are really getting to me," he said.

They walked across the bridge from the island to the opposite bank of the Sulphur River. The sun was shining brightly and it was nice to breathe in some fresh air again. A path trailed in front of them, leading into some woods.

Tom wondered if it was the same wood as before. Maybe they were close to the end of the adventure. He spotted a computer screen flickering ahead of them. It was cut into the trunk of one of the trees close to the path. He rushed over and inspected the picture. There was his bedroom as before, but there was yet another clue letter.

'PASSWORD . . . M E H I C A S,' flashed in front of him.

Tom couldn't think of a single word containing those letters. Zoe came to his rescue though.

"Christmas," she said.

"That's no good," said Tom. "You haven't used the E."

He thought for a while longer and then came up with 'MECHANICS.'

'ILLEGAL PASSWORD . . . M E H I C A S.'

"Come on, let's carry on," said Tom. He was confident that they could solve anything that the Crusade threw at them. At least, he was sure that he wouldn't be able to guess the password and he didn't want to waste any more time trying.

They continued to walk along the path and eventually they came to a clearing. Within the clearing the path sank downwards so that it was almost like a trench. At the end of the trench was a door placed into the side of a hill. There was a notice on the door. It said, 'RING BELL TO ENTER' and there was an arrow pointing to the bell-push. They could see the bell above the door, quite high up. Tom boldly pushed the button but the bell didn't ring. He tried pushing harder and for longer but it still didn't work.

"We could knock," said Zoe.

Tom rapped the wooden door hard with his knuckles. There was no answer.

Chapter Eight - The Ring of the Bell

After two minutes of thumping on the door and receiving no answer, Tom thought that maybe the problem had to be solved in some other way.

"We need to ring the bell," he said, stating the obvious and looking up at the electric bell above his head.

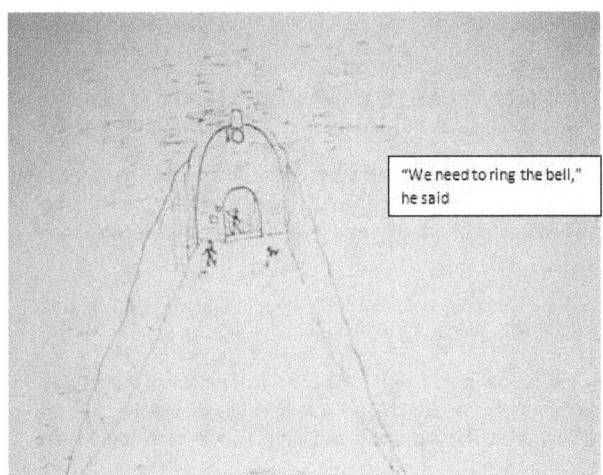

"We need to ring the bell," he said

"But it doesn't work," said Zoe, who was trying the bell-push for herself. "It must be broken."

"So we need to find out why it's not working and then fix it," said Tom, stepping back a bit to assess the situation from a new angle.

He tried to sound confident but he didn't really know what to do. How could he fix a piece of electrical equipment? He wasn't an electrician and his father had always warned him not to fiddle around with electrical things because it could be very dangerous.

"I'll go and have a closer look at the bell," he said. "You stay here and push the button when I say. Don't try it until then because I might get electrocuted. Do you understand, Zak?"

"Yes, I understand," said Zoe, and she sat down on the doorstep with Grizz.

Tom walked back along the trench that led to the door. He was determined to sort out this problem if he could. Zoe had done so well at the Sulphur River and he felt that it was time for him to do some of the thinking. When he got to level ground he stepped to the left of the path and then climbed the steep ridge up to the doorbell. It was quite an effort to scramble up the last few metres as the rock was slippery, but eventually he made it to the doorbell. The bell was massive but otherwise it was just like the electric bell that they had at home.

Tom knew that electrical things sometimes didn't work because a wire had become loose. Without touching the bell he looked for any sign of a wire that was not connected properly. He could not see any obvious fault. As far as he could see there was nothing wrong with the bell at all. He leaned over and looked down at Zoe below him. She was leaning against the door and squinting up at him.

"Try the button," said Tom, at least to give the impression that he had done something up there and that he knew what he was doing.

Zoe pushed and Tom carefully observed the bell. Nothing happened at all. The hammer of the bell remained absolutely still. There were no sparks around the contacts or any other sign that electricity was flowing in the coils of the wire. Sometimes wire heated up a bit when electricity was running through it, but Tom was sure that no electricity was getting to the bell at all.

"It's not working," he called, immediately realising that this would be obvious to Zoe. Before she could comment he said, "So I'll just check the connections. Stop pushing the button."

Tom looked at the connections to the bell. They were thick, metal strips made of copper and attached to the bell on either side. One strip went to the left and one went to the right. Each one was about five centimetres wide and one centimetre thick and they both looked shiny and bright in the sunlight.

Tom wondered why the connecting wires were so thick. He thought that maybe it was because the bell took a lot of electricity to make it work because it was so big.

Anyway, he decided to follow the copper strip around. He knew that electrical things would not work if there was a gap anywhere in the circuit. This was because electricity could not jump across a gap so it didn't flow at all. If there was a gap in the copper strip somewhere then this would stop the bell from working.

He decided to follow the strip connected to the right side of the bell because that was closer to him. It seemed to be slotted into a groove in the rock of the ridge up which he had just clambered. It was secured to the rock with bolts every metre or so. Without touching it, he followed it with his eyes as he walked back down to ground level. Just as he got to the bottom of the ridge, the copper strip stopped. This was obviously why the bell couldn't work. There was no complete circuit for the electricity. He walked back up the ridge to the bell.

"Tom, what are you doing?" called Zoe as he got back to the large bell.

"I'm checking the connections," said Tom, importantly. "I'll be down in a minute."

"Do you want me to push the button again?" said Zoe.

"No! Whatever you do, don't push the button. Just wait there for a moment."

Tom saw that there was a narrow ledge upon which the bell was mounted. There was just about room for him to squeeze along it so that he could have a closer look at the connecting strip. As he moved past the bell his arm caught against the hammer. He almost overbalanced and fell down into the trench, but he just managed to grab metal connection on the left side of the bell. He hoped that it wasn't live with electricity but the choice was electrocution or falling a few metres onto solid rock. Fortunately, the metal connection was not live, but as he pulled himself back onto the ledge, the hammer sprang back and hit the gong of the bell.

'Ding,' it went.

"Have you mended it, Tom?" cried Zoe from below.

"No, I haven't," said Tom, with annoyance. "But I may have found a way of opening the door."

He leant over and grabbed the hammer again. He pulled it over and then let go.

'Ding,' went the bell.

He repeated the operation several times as quickly as he could.

'Ding, ding, ding, ding,' went the bell.

"Try the door," shouted Tom, breathless with the effort of pulling the large bell hammer.

Zoe twisted the door knob and pushed.

"It won't budge," she said.

'Ding, ding, ding, ding,' went the bell as Tom tried his trick again.

"Pull it," he shouted. "Hard!"

Zoe twisted the handle and pulled as hard as she could. The door knob came loose out of the door and she fell backwards. The door remained shut but the door knob, attached to a length of metal rod, was now lying on the ground next to Zoe.

Tom saw all this from his vantage point above.

"You idiot, Zak," he said. "You've gone and broken it now. We'll never get in."

"You're the idiot, Tom," shouted Zoe, lying on her back and looking up at her brother. "You told me to pull it and that's what I did. Anyway, what do you think you're doing up there? I wouldn't be surprised if you've broken the bell."

Tom knew that Zoe was right. He had been a fool. Once again, he was taking his anger out on her because he hadn't solved the problem and he didn't like to look a fool.

"Sorry, Zak," he called down, softly.

He decided to follow the copper strip down to ground level on the other side of the trench. It, too, was bolted to the rock within a shallow groove. The first thing that he noticed was that the strip did not follow the ridge immediately but went down to the bell-push by the door, and then it came back up to where he was standing. Then the strip followed the ridge down on the other side of the trench. He walked beside it but about halfway down he was obstructed by a large sealed box about the size of a refrigerator. The copper strip was connected to one side of the box which was labelled with a plus sign. The strip reappeared from the opposite side of the box which was labelled with a minus sign.

"It's a huge battery," said Tom to himself. He skirted around behind the battery and continued on down the slope. He carried on following the shiny strip of copper and then it stopped, just like it had done on the other side of the trench. The interesting thing was that the strip stopped exactly opposite the other piece.

"This is the gap in the circuit," mumbled Tom. "And this is why the bell won't work."

The gap was the width of the pathway which was about a metre or so at that point. Tom went back to Zoe to tell her the news.

"There's a gap in the electrical circuit," he said. "Look, I'll show you."

He held out his hand and pulled Zoe to her feet. She was still lying on the ground in a little patch of shade formed by the walls of the trench. Tom was grateful for the chance to make up with Zoe after their latest disagreement.

Tom led Zoe back along the trench to the point where the copper connections stopped. About waist high on either side was a gap in the trench, and they could quite clearly see the bare metal connections.

"We need a wire to cross from here to here," said Tom, pointing to the ends of the copper strips. "Then the electrical circuit will be complete and the bell will ring when we push the button."

"And I found the connector," said Zoe, giggling. "It came off in my hand."

"Of course!" said Tom, excitedly. He realised that Zoe was referring to the length of metal that she had pulled out of the door when she had tried to open it.

He rushed back to the door and picked up the piece of metal. Tom thought that it was clearly designed for the gap as it looked to be exactly the right length. Once again, they were about to solve a level in the Crusade.

Tom was smiling broadly when he got back to Zoe with the connector.

"Well done, Zak," he said. "I'm sorry I shouted at you before."

He placed one end of the metal connector into the slot on one side of the trench. He pushed it up against the copper strip to make sure of a good connection. He then placed the other end into the opposite slot but, to his horror, he saw that it didn't fit. It was too short by about a centimetre. There was still a gap in the circuit. It wasn't a very big gap and it wasn't as big a gap as before, but it was still a gap. The electricity would not be able to jump the gap and so the bell could not work.

"It's too short," he said out loud so that Zoe could hear. "Look, Zak, it's a tiny bit too short."

He sat down with Grizz in the shade and started to turn out his pockets.

"What are you looking for?" said Zoe.

"Something metal to close the gap," said Tom. "But I don't think I've got anything. Everything is made of plastic or cloth. Even the zip on my jacket and Grizz's lead clip are made of plastic."

Zoe checked her pockets too, but she didn't have anything metal either. She sat down beside Tom in the only bit of shade left in the trench. It really was getting hot now. The sun was shining fiercely and there was not a cloud in the sky. Grizz lay between them, panting, his tongue hanging out in an attempt to keep cool.

Within a few minutes their little bit of shade had disappeared completely as the sun rose higher into the sky.

"It's so hot," said Zoe, beads of sweat on her forehead.

"It certainly is," said Tom. "The mercury will be rising today."

"Whatever do you mean now, Tom?" said Zoe.

"I mean it's hot," said Tom. He didn't have the energy to explain himself more fully. "I think we'll have to move into the shade of the trees."

"Yes, OK," said Zoe, getting to her feet, and they walked towards the coolness of the trees.

"Are you talking about the planet Mercury?" said Zoe as they got to the edge of the woods.

"No," said Tom. "I meant the liquid inside a thermometer. It's called mercury."

"So, what do you mean when you say that the mercury is rising?" said Zoe.

They sat down and leant against the trunk of a large oak tree. It was refreshing to be out of the direct heat of the sun. Tom sighed. He was annoyed with the Crusade because he was sure that they had solved it. Perhaps the Crusade was impossible after all.

"Well?" said Zoe, interrupting his thoughts.

"Well, what?" said Tom, confused. He had been trying so hard to think of a solution to the circuit problem that he had not really been listening to his sister.

"The mercury will be rising. What does it mean?"

"It means that when the thermometer gets hot the mercury expands or takes up more space. So it rises up the tube of the thermometer. That's how you can tell what temperature it is."

"Does everything expand when it gets hot?" said Zoe.

"More or less," said Tom.

"So the connector will expand too," said Zoe. "In this heat it might have stretched across the gap while we've been sitting here."

"You're right, Zak," said Tom, leaping to his feet with renewed energy. "Quick, let's go before the sun starts to set and the temperature goes back down."

They ran back along the path to the trench even though it was really too hot to run. Tom got to the metal connector first. His heart was pounding with the run and with the excitement of what he hoped to see. But, to his despair, he saw that the gap was still about the same as before.

"It's no good," he said, as Zoe came over to see for herself. "It needs to get a lot, lot hotter for it to expand enough to cross the gap. It's hardly got any bigger at all."

"Couldn't we build a fire underneath it?" said Zoe. "That will make it really hot."

"Right, Zak. It's worth a try, isn't it? Let's collect some twigs and branches and build one. Make sure that the wood is really dry."

They both wandered off into the trees to collect firewood. It didn't take long even though they were both hot and tired. There seemed to be a lot of dead wood around. In a few minutes they had enough wood to go right along the width of the trench, underneath the metal connector. Tom arranged the pieces so that there were enough small twigs to get the fire going, and then some larger pieces to give out all the heat.

""All we need to do now is light it," he said when he had finished arranging the last few logs. His heart sank a little when he realised what he had just said. "Anybody got any matches?"

Zoe was not prepared to give up.

"Rub two sticks together," she said.

She already had two small sticks in her hands and she began rubbing them as quickly as she could. Tom tried too but, before long, it became quite clear that it wasn't going to work. There was obviously a skill to it that neither of them had ever learnt. The effort of trying just made them sweat even more.

"Perhaps we've got the wrong connector," said Tom. "Maybe we made a mistake by pulling the door handle out of the door." He was beginning to despair.

"But why else would the door handle be so long?" said Zoe. "It has to be the right connector. We just have to think of a way of lighting the fire."

At that point Tom remembered the magnifying glass. He had long since taken off his jacket but, in the pocket, he found the small, round piece of glass.

"This is the answer," he said, triumphantly, holding the lens between his thumb and forefinger. "This will light the fire for us."

Tom knelt down and used the magnifying glass to focus the energy of the sun into a single, bright point. He directed the spot of light onto a dry piece of moss and it soon began to smoulder. A flame flickered for a moment and then some twigs caught alight. The wood was very dry and in seconds the whole fire was ablaze. The heat was incredible and the three of them had to shrink back from it. Soon they were right up against the locked door and still the fire got fiercer.

Soon they were right up to the locked door and still the fire got more fierce.

"If this doesn't work quickly we'll be roasted alive," shouted Zoe above the roar and crackle of the flames.

The metal connecting bar was now hidden by the bright, yellow-white glare of the fire.

"Try the bell-push now," shouted Tom as the fire spat out bullets of red-hot embers. "The bar must have expanded by now."

Zoe pushed the button and the bell rang out loudly.

"It's the fire-brigade," she said with delight, laughing at her own joke.

As she said it, the door opened behind them and they fell backwards into the hillside. Grizz jumped in after them and the door, which was on a heavy spring, slammed shut. It was suddenly very quiet. After the roar of the fire and the ringing of the bell, everything had become very calm. Tom got up and dusted himself down. On the wall beside the door was the familiar computer screen. He rushed over to see if there was anything new. There was another letter, the eighth one altogether.

The screen flashed, 'PASSWORD . . . M E H I C A S E.'

'MECHANISE,' he typed and pressed RETURN.

'ILLEGAL PASSWORD . . . M E H I C A S E.'

He racked his brains for another possible word containing the eight letters but was disturbed when Zoe grabbed his arm.

He turned round to see himself reflected in a huge mirror a few metres away. He was standing next to Grizz and Zoe, but also to be seen in the mirror was the white, ghostly figure of a man.

The figure was talking to them.

"Welcome to the Mirror Maze," he said in a hoarse whisper. "Remember, right is wrong and left is right."

Chapter Nine - The Mirror Maze

"Right is wrong and left is right."

The figure then disappeared and Tom, Zoe and Grizz were left staring at themselves in the mirror.

"Did you see that?" said Tom, quietly.

"Did you hear that?" whispered Zoe. "That was spooky. And what does it mean? Right is wrong and left is right. It doesn't make sense."

"I'm not sure," said Tom. "Anyway, let's go," he continued, not wishing to waste any more time.

They walked up to the mirror and met their reflections close up. There was a choice at the mirror. There was a turn to the left or a turn to the right, and both exits led into dimly-lit tunnels.

"Right is wrong and left is right," muttered Tom to himself, and he led the way by turning to his left.

"Left is right," he said loudly and confidently, and off he went into the left-hand tunnel. Zoe and Grizz followed obediently.

The tunnel had many twists and turns and was also quite low in places. It was tiring having to stoop so low and still walk along. Tom felt his legs start to ache. Eventually, they turned a bend and saw some figures in the distance.

"Hallo!" shouted Tom.

"Hallo," came the reply a moment later.

"At least they're friendly," said Tom.

As they got closer, Tom realised that the figures were reflections of themselves.

"It's us," he said.

"I can see that," said Zoe. "But how come they talked to us?"

"It was the echo of my voice," said Tom. "Listen."

He shouted out again. "Hallo!"

"Hallo," replied the echo.

"Sound energy bounces off the walls of the tunnel and comes back to us," said Tom.

By now they were much nearer the mirror and their reflections were quite close. When they were about twenty paces away, the same ghostly figure reappeared. It turned to face their reflections.

"Right is wrong and left is right," it said, and then it disappeared.

At the mirror they faced the same two-way decision as before. There was a left turn and a right turn.

"We must turn left again," said Tom. "Left is right," he whispered, hoarsely, trying to imitate the ghost.

"I don't like this," said Zoe. "I've got a feeling that left is not right, but I don't know why."

"Well, left is left, of course," said Tom, enjoying the play on words. "But left is right, so we'd better turn left."

He led the way down the left tunnel.

It wasn't long before they approached another mirror in the maze, but by now Tom's legs were feeling very weary. He noticed that Zoe had gone very quiet and that Grizz was even quieter, almost mournful. The ghost appeared in the mirror when they got close.

"Do you mean left is correct?" blurted out Zoe, before the ghost had a chance to speak.

Tom saw her image mouthing the words in the mirror and then the ghost turned towards her image to reply.

"I do, my dear," it said, in a croaky whisper. "Left is right. Right is wrong." Then it disappeared.

"What if it's lying, Tom? I don't like this at all. My legs feel funny," said Zoe.

"Look, Zak," said Tom. "The Crusade doesn't lie. It never has, not on the whole adventure. So why should it start lying now?"

At the mirror he turned to the left leaving Zoe and Grizz to chase after him. He couldn't help noticing the worried look on Zoe's face as he looked at her reflection, but he thought it best if they carried on. The Crusade was obviously testing their will to carry on.

Tom had to wait quite a long time when he got to the next mirror. He leant against the wall because he was so tired, but he was also annoyed at Zoe because she was dawdling so much and he wanted to show his annoyance when she finally appeared.

She eventually came staggering around the last bend with Grizz tripping along behind.

"Tom," she said, weakly, "look at your reflection."

Tom saw that the lower part of his body was very hazy in the mirror. He also noticed that the only part of Zoe that was clear was her head, and that Grizz was merely a shadow. The ghost appeared beside them in the mirror. It turned towards their own ghost-like forms.

"Left is right and right is wrong," it said, rather louder than before.

Tom also saw that the reflection of the ghost seemed stronger and clearer than before. It was also a lot clearer than their own reflections, yet the ghost did not exist beside them in reality. He didn't feel at all well either, but could not understand why anything was wrong, unless this was just a test of their stamina.

He scratched his head, trying to think. He looked at his faint reflection and saw his transparent legs, his red jacket, the frown on his face and his left hand scratching his head in thought.

But he wasn't scratching his head with his left hand, he thought. He was scratching it with his right hand. Of course, the mirror image was reversed. The ghost only existed in the mirror and always spoke to their reflections. So when it told them to turn left, it meant their reflections had to turn left. And the only way to get their reflections to turn to the left was for themselves to turn to the right. They had been taking the wrong option every single time.

"Quick, Zoe, Grizz," said Tom, urgently. "We've got to go back before our reflections disappear. If we have no reflection then we cannot exist. If we don't exist then we've lost the game."

He turned and pulled Zoe along with him.

"Grizz, come on, boy," he called, encouraging the little dog to follow.

Back they went, tracing their exact route until they got back to the door with the bell and the computer monitor.

They stood in front of the first mirror again. Their reflections were solid and true. The ghostly form appeared before them, just as before.

"Welcome to the Mirror Maze," it said. "Remember, right is wrong and left is right."

"Look," said Tom, "the ghost talks to our reflections, not to us. We have to make our reflections turn to the left."

They walked towards the mirror and watched their mirror images very carefully. As they reached the junction, they turned to the right. Their reflections turned to the left.

"So right is right, not wrong," said Zoe, cheerfully.

After a couple of bends in the tunnel, they entered a cave which was glowing with an eerie, blue light. Fairground music was blaring in their ears and there was a general bustle of activity. There were creatures of all kinds milling around. Some were recognisable but others were completely alien. Some were big and some were small. All in all, they made a very strange crowd.

Suddenly, a very odd-looking fellow jumped out in front of them, waving a fistful of papers.

"Roll up, roll up, get your tickets here," he cried in a shrill voice.

"No, thank you, we haven't any money," said Tom, politely but firmly.

"But you don't have a choice," said the ticket-man. "Besides, they don't cost anything."

He was a strange chap to be sure. He was clean-shaven but he had a mop of frizzy, blue hair and piercing, black eyes. He stooped forward because of the weight of the tray of tickets that he was trying to get rid of, and, before Tom could object, he clipped a piece of paper to Zoe's jacket. He did the same to Tom and then he fastened another to Grizz's collar.

At that moment, the eerie light turned from blue to green and this gave them all a rather sick-looking complexion.

"Come along, hurry, hurry," said the weird character. "Join the queue."

He ushered them along until they were standing in a long line of creatures. They all had similar tickets attached to their clothing, fur or tentacles. There was certainly a great variety of creatures in the queue and most of them had terrified expressions on their faces.

The ticket seller then wandered off into the crowd looking for more business.

"Roll up, roll up for the Black Death Slide," he called, and before Tom could ask what the tickets were for, the man had disappeared.

"Roll up, roll up for the Black Death Slide."

At that moment, the sickly green light changed to red and everything around them took on a fiery glow. They found themselves being herded along in a queue between wire mesh barriers. There were creatures in front and now there were some behind, and steadily the line was edging forward. Everyone in the queue had a ticket attached to them for the same fairground ride.

In the red glow, Tom could see that the queue went around in a huge circle and they were high up on the inside of a massive dome. It was a bit like the dome inside St. Paul's Cathedral. He could see the front of the queue which was opposite where they were standing at that moment. The creature at the front of the line then sat down on a slide and disappeared downwards. There were about thirty playground slides to choose from and each one disappeared into a misty cloud in the centre of the dome. The cloud was glowing red in the strange light and then it glowed blue as the lighting changed again.

Above the racket of the organ music they could still hear the scream of the poor creature that had just disappeared into the cloud. The next creature in the line then stepped forward and selected one of the slides. The slides were coloured blue and black and were made of a shiny, plastic material. Even from where they were standing, Tom and Zoe could see the look of terror on the dwarf-like features of the little person on the slide. He let go and slid steeply downwards at great speed into the blue cloud. A high pitched scream pierced the air.

Zoe looked over her shoulder at Tom as they trudged slowly towards the head of the line. As she did so, the light switched from blue to green, which only seemed to exaggerate the fearful look on her face.

Tom took a closer look at the ticket clipped to Zoe's jacket. The words 'Black Death Slide' were printed across the top. Then there was a number, '59568' and underneath were the words 'Black Slide Safe.' He looked at his own ticket which was attached to the collar of his jacket. It was the same as Zoe's except for one detail. The number of his ticket was '59569.'

"The black slide is safe, Zak," he said to Zoe. "When it's your turn, take the black slide."

Zoe looked with Tom over towards the row of slides. Some were green and some were black. Then the dome lighting changed from green to red and some of the slides changed colour as well. Some changed from green to black and some changed from black to red.

"Which black slide?" said Zoe.

It was the obvious question to ask but for the moment Tom had no answer. There was more than one black slide and, not only that, the slides were changing colour as well. Once again, it seemed as though the Crusade was playing with their lives, and Tom thought it was unfair.

It was difficult to concentrate with all the music and screaming. Tom looked again at his ticket.

"Black slide safe," he said out loud.

At that instant, the lighting changed again, this time from red to blue. He noticed that his jacket changed colour as well. It changed from red to black.

"How odd," he said to himself. "I wonder why that happened."

By now there were just forty other creatures between them and the front of the queue. There was no way of escaping from the line. They were surrounded by chicken-wire mesh on both sides and above their heads. They had no choice but to step forward in the line because they were being pushed from behind by the next creature in the queue.

Tom could see that there was an usher at the front, similar in appearance to the ticket seller. The usher prodded each creature at the head of the line so that they stepped forward.

"Make your choice," he would shout, dramatically, and the poor creature would make a dash for one of the slides and jump.

A hairy, six-legged animal refused to make a choice when instructed. It stood there, rooted to the spot, all six knees trembling. The poor, frightened beast was then grabbed by a crane as though it was a cuddly toy at a funfair arcade. It was lifted into the air, swivelled around and dropped over the side anyway. It disappeared with a dreadful howling into the misty, blue cloud.

"You might as well make a choice," shouted the usher to everyone else in the line. "Otherwise you die anyway."

The light changed from blue to green. Tom noted with interest that his jacket remained black in colour, and he had a feeling that his brain was beginning to solve the problem.

"Zoe!" he said, urgently. "What colour is my jacket normally?"

"Bright red," said Zoe, "and it's awful."

"Well, never mind how fashionable it is," said Tom, a little offended. "It looks red because in white daylight it reflects red light and absorbs all the other colours. White light contains all the colours mixed together. My red jacket can only reflect red light and that's why it looks red."

"So what?" said Zoe. "I'm more worried about getting to the front of the line. There are only thirty-three ahead of me now."

"Thirty-two, actually," said Tom, as a fluffy ball of fur hurtled down one of the black slides. Its eyes were bulging outwards and it screamed as it entered the misty, green cloud.

"My jacket looks black, doesn't it?" said Tom, persisting with his idea. "That's because there is no red light to reflect. It can't reflect green light so it doesn't reflect any light at all. An object that doesn't reflect any light looks black."

"So how does that help us?" said Zoe as the dome lighting changed to red and Tom's jacket turned from black to red.

"Look, my jacket is red now. It can reflect the red light so it looks red," said Tom, excitedly. "Now, look at the slides. I reckon that only one of them is really black in colour."

"But there are at least ten black slides, Tom. Look," said Zoe.

"Listen," said Tom. "One slide is black. The others are either red, blue or green. In red light, only the red slides will look red because they can reflect red light. All the others will look black."

They looked over at the row of thirty slides. In different circumstances, it might have been good fun to slide down one of them.

"Make your choice," boomed the voice of the frizzy-haired usher.

The next unfortunate creature selected one of the black-looking slides, sat down and let go. As it sped downwards, the light changed to blue and so did the slide upon which it was sitting. It fell screaming into the misty, blue cloud. Out of the thirty slides, about were now coloured blue. The rest were black.

"So how does all this help us?" shouted Zoe, above the screams. There was an air of panic about the creatures ahead of them in the queue and it was not easy for Tom to hear what she was saying.

"We need to spot the slide that never changes colour whatever the colour of the light. There must be one that really is black. Black objects look black because they never reflect any colour of light."

"So how does that help us," shouted Zoe again.

Tom could tell that she was getting caught up in the panic as well, but he tried to remain calm. Time was running out but he had to be sure that Zoe understood his plan.

"The dome lighting will probably change about ten more times before we have to choose one of the slides. You look at the first six slides and I'll look at the next six. We need to watch for the slide that stays black in red, blue and green light."

They put the plan into action immediately. They were getting closer and closer to the front as they trudged steadily forward, Zoe first, then Tom and Grizz behind.

After the three colour changes Tom had seen that his six slides had all changed colour at some stage.

"All mine changed colour!" shouted Zoe.

"Look at number thirteen to eighteen then," shouted Tom in reply.

At the end of the next three colour changes, Tom could scarcely contain his excitement.

"It's the twenty-second one!" he yelled. "It stayed black all the time."

The discovery was made in the nick of time. Zoe was at the head of the queue at that moment.

""Make your choice," screeched the usher, taking Zoe's ticket and prodding her forward. She seemed to hesitate and the crane came over to grab her.

"Run!" shouted Tom, but Zoe was already running. She sped over to slide number twenty-two and jumped. Tom heard her screaming as she disappeared into the misty cloud.

"Make your choice," shouted the usher, looking at Tom and ripping the ticket from his clothing. Tom did the same as Zoe. He ran over to slide twenty-two, sat down and let himself go. As he went over the edge he looked back at Grizz. Their eyes met and Tom hoped that Grizz understood enough to follow his master down the same route.

Through the cloud he sped and then suddenly he was out into bright sunshine, still sitting on the slide and all was strangely quiet. He looked to either side of him and saw the other slides emerging from the cloud as well. Some were red, some were blue and some were green. All of them came to an end directly below the cloud leaving a drop into thin air. The only slide that continued onwards was the black one that he was sitting on. After a while it flattened out a bit and Tom felt himself slowing down. The slide continued almost to ground level where it entered an opening in the side of a grey-stoned building.

He finally landed with a slight bump onto a bouncy floor made of some sort of rubber. Next to him was Zoe and, to his delight, a few seconds later, Grizz came hurtling in to land between them.

"That was fun," said Tom, laughing. His laugh was a nervous one though because it hadn't really been fun at all. It was just a relief to be with Zoe and Grizz again. The experience had obviously affected him in some way because he was tingling all over.

Chapter Ten - Grizz the Frizz

They were in a massive hall just like one of the main railway stations in London. It was lit by windows high above them in the roof and the frames cast symmetrical patterns on the rubbery floor. They could see that they had entered the building through a hatch like a large cat-flap in a door. Tom pushed the flap open a little and looked outside. He could see the long, black slide disappearing up into the clouds. It had been quite an exciting ride to say the least and perhaps a little frightening too.

Tom allowed the flap to drop back down and turned his attention to the computer monitor that was placed in the wall beside them. As he scrambled over to it, he was disturbed by Zoe's laughter.

"Tom, your hair," she managed to say between giggles. "And look at Grizz."

Tom looked at the dog and saw that his fur was sticking up on end.

"It's Grizz the Frizz!" he said, laughing out loud.

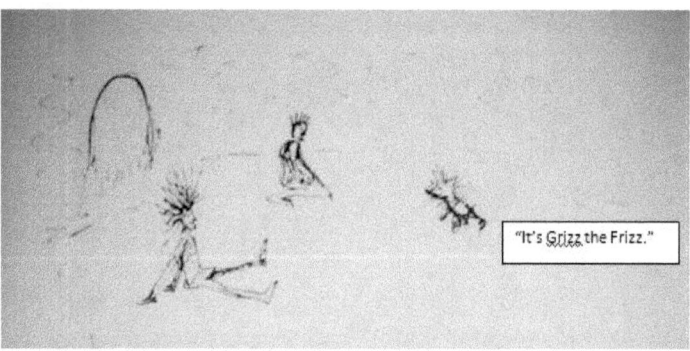

"It's Grizz the Frizz."

He also noticed that Zoe's hair was sticking up like the spines on a hedgehog.

"And you're just as bad, Zak, if not worse."

"What's happened?" said Zoe, trying to brush her hair flat with her hand. Every time she let go, her hair immediately sprang back up.

"We're charged up," said Tom. "It must have been because of the ride down the plastic slide. The friction charges us up. It's just like when you rub a plastic comb against your jumper and you can pick up little pieces of paper with it."

"Is it the same as when you rub a balloon and it sticks to the wall?" said Zoe.

"Yes, that's right," said Tom, quite impressed by his sister's knowledge again. "It's called static electricity."

"So, how can we get uncharged?" said Zoe. "I'm not too keen on wandering around looking like this."

"You get discharged, not uncharged," said Tom. "And I think it's an improvement."

"How do we get discharged?" said Zoe, flattening her hair with both arms.

"It's easy," said Tom. "You can do it by touching something metal and then the static electricity can run away."

He looked around for something metal that he could discharge himself with. It was then that he noticed that there was nothing metal around. The floor was made of some kind of rubber and the walls were coated with a plastic material. He needed a substance that would conduct electricity and both rubber and plastic were insulators. Only conductors could take away the charge and there were not any around. It was all rather odd.

"Oh well," he said. "It doesn't really matter. Let's not worry about it just now. It won't hurt you as long as you don't mind looking like you've been pulled through a hedge backwards. Personally, I think it suits you."

"No, it doesn't Tom," said Zoe, looking rather hurt.

"I'm only teasing, Zak. Come and have a look at this."

Tom was peering at the computer screen. The same familiar bedroom scene was shown but there was an extra clue to the password.

'M E H I C A S E D,' flashed at the bottom of the screen. They had now earned a ninth clue letter.

Tom typed 'MECHANISED,' onto the plastic keyboard but had the feeling that he might have tried it before at an earlier level. He pressed RETURN with little hope of success.

'ILLEGAL PASSWORD . . . M E H I C A S E D,' came the inevitable reply.

"I can't believe that the password is so long and complicated," he said to himself.

Meanwhile, Zoe was busy trying to flatten Grizz's fur but, like her own hair, it kept springing back up again.

"Look!" she said, laughing loudly. "Grizz's ears are sticking on end too. And his tail!"

Tom laughed too. Grizz looked like a cartoon character that had seen a ghost.

At that moment, their attention was distracted by an announcement booming loudly from some hidden loudspeaker system.

"The jet-bug for Hitchin's Wood might be leaving in five minutes from platform seventeen," it said, and the announcement echoed through the building.

They looked up and, in the far distance, they noticed some sort of vehicle glide into the hall and come to a halt. A sign hanging from wires above them lit up brightly. It said 'Platform Seventeen. Hitchin's Wood.'

"Quick! Let's go," said Tom excitedly. "It must be the end of the Crusade. We've done it, Zak."

They rushed across the hall, bouncing on the rubber surface. They passed platforms one to sixteen and saw other jet-bugs waiting to pick up passengers. There were all sorts of other destinations shown on the display boards. There was Dragon's Lair, Cuddly Corner and Sore Point, to name but three.

They sped onto platform seventeen and approached the jet-bug. It was now stationary but had entered the building along metal rails that stretched out in front of them into the open air. The jet-bug was like a train but had only a single carriage. There was an illuminated sign on the front which read 'Hitchin's Wood.' As they walked up to the vehicle, the doors slid open silently.

At that moment there was a blinding flash and a deafening explosion. It came from further along the hall, probably about platform fifty, Tom guessed, but he couldn't be sure. To be honest, he wasn't too concerned as long as no person, or no creature, had been hurt. Without investigating any further, they stepped into the interior of the jet-bug.

It had plush, velvet covered seats and there was even thick-pile carpet on the floor. There wasn't a driver but there was a control panel at the front which had various switches, buttons and lights.

Tom sat down and relaxed for the first time in ages. He nestled down into the seat, closed his eyes and sighed.

"Luxury," he said out loud, leaning his head against the head-rest. His hair was sticking up and he felt a bit odd, but he didn't care. They were on their way home at last.

He could quite easily have dozed off but was startled by a beeping sound coming from the front of the buggy. He opened his eyes and spotted a flashing, red light on the control panel. Zoe was already investigating.

"We're out of fuel," she said.

Tom went over and saw the warning message for himself. Written on the flashing light were the words, 'Out of fuel' and the beeping noise only served to annoy him further.

Almost immediately there came another announcement over the station's public address system.

"The jet-bug at platform seventeen has been delayed because it has run out of fuel," it said.

Zoe heard the announcement too. She had been trying to control her hair by holding it in a pony tail but, when she heard the news, she let go and her hair sprang back into life. At any other time it would have been comical but neither of them found it funny.

There then followed another explosion. This one was much closer and it rocked the jet-bug back and forth. Tom stepped back onto the platform to investigate. The jet-bug that had been at platform eighteen had been blown to pieces. It was quite a tidy explosion because there didn't appear to be any pieces lying around. Otherwise it might have been extremely dangerous but the jet-bug had simply disappeared in a puff of smoke.

There were two further announcements.

"The jet-bug for Cavalier Court might leave from platform twenty-three in five minutes." That was followed by, "The jet-bug for Slime Square, which was at platform eighteen, has been cancelled."

"Well, that's obvious," said Tom, under his breath, and looking at the deserted platform next to him. "It's disappeared."

He went back into his own jet-bug, the one that was destined for Hitchin's Wood and home. The fuel light was still flashing and Zoe was sitting mournfully with Grizz on her lap. They were a very sorry sight.

"What's going on, Tom?" said Zoe.

"I'm not sure," said Tom. "The buggy on the next platform just disappeared. What we need to do is find out how to refuel."

He went back to the front of the vehicle to look for clues. It was then that he noticed the two shiny, metal fuel pumps just outside the front window. He couldn't think why he hadn't noticed them before but he was no longer too surprised by the Crusade.

"Look, Zoe, Grizz, the petrol pumps are out there. Perhaps it's self-service. Maybe we need to refuel the jet-bugs ourselves."

Zoe came to the front carrying the fluffy, frizzy ball of fur that was Grizz, and looked out the window to where Tom had indicated.

There was then another loud explosion from nearby. This was followed by yet another announcement as soon as the echo had died down.

"The jet-bug at platform twenty-three for Cavalier Court has been delayed because it has run out of fuel.

And this was followed by, "The jet-bug for Guppy Creek that was at platform seventy-one has been cancelled."

"None of the buggies seem to have any fuel," said Tom. "And it seems as though refuelling can be quite a dangerous operation."

"How do you know that?" said Zoe.

"I don't. I'm just guessing," said Tom. "But the announcements keep saying there's no fuel and then there's an explosion."

"I know petrol can be very dangerous," said Zoe. "That's why there are 'No Smoking' signs in the petrol stations."

"That's right," said Tom. "Petrol is very flammable. It's also why there is no smoking allowed on planes when they take-off and land. Aircraft fuel is even more likely to explode."

"Do you think the jet-bugs are exploding because the fuel has caught fire?" said Zoe.

"Yes," said Tom, "but it seems a bit odd that it's happening so much."

At that instant, there was another massive explosion. Tom looked outside to see if there was anything happening. Once again, there was no sign of any flying pieces from the explosion and all seemed quiet. He studied the fuel pumps beside their jet-bug and rubbed his hands through his hair, deep in thought. His hair still felt rather strange. If only he could discharge the static electricity in him.

He then realised what must be happening. All travellers in the jet-bug station must be there because of the Crusade. They must have solved the problems just like they had, and had finally arrived her by sliding down the Black Death Slide. Because the slide was made of plastic, they had all charged up with electricity. To get rid of the electricity, something metal was needed and the fuel pumps were made of metal. But touching the fuel pump was a mistake because the electricity jumped across to the fuel pump with a spark. The spark ignited the fuel and that was the reason for the explosions. End of game.

"We've got another problem to solve," said Tom to Zoe. "We can only leave here if we refuel the jet-bug."

"So let's do it," said Zoe. "What's the problem?"

"The problem is that we're charged up with electricity. If either of us, or Grizz for that matter, goes near the fuel pumps we will cause a spark and the fuel will explode. And that means the end of the game."

"I should think that means the end of everything," said Zoe. "So what can we do?"

"We need to get rid of the static electricity before we try to refuel. Any ideas?" said Tom, hopefully.

"No, not yet," said Zoe. "But I'm going to solve this somehow."

She walked out of the jet-bug and onto the platform.

"Don't go near the fuel pumps," shouted Tom, hanging onto Grizz in case the little dog decided to chase after Zoe.

Zoe went over to the gleaming pumps but did as Tom had said. She went no closer than was necessary to read the writing on the pumps. Tom saw her laughing to herself as she did so.

There was then another loud explosion from along the hall and it shook the building. Zoe rushed back into the comfort of the jet-bug.

"Have you thought of anything?" said Tom with a hint of desperation in his voice.

"Why didn't Oxide score anything?" said Zoe.

"Why didn't oxide score anything?" said Zoe.

"What are you going on about, Zak? Have you finally cracked? I'm trying to get us out of here and you're talking gibberish."

"It's written on the pump," said Zoe.

"What is?" said Tom.

"The score. The result of some netball game or something."

"What did it say exactly?" said Tom.

"Carbon eight, hydrogen eighteen on one pump. And hydrogen two, oxide . . . and there's no score, on the other pump."

"Was there anything else written on the pumps?" said Tom, confused by what Zoe was saying.

"Yes, octagon, I think," said Zoe. "At least, it was oct something."

Tom went outside to read the words for himself. Seconds later he shouted to Zoe to join him on the platform.

"Stand there!" he said, pointing to a spot in front of the second pump. "And hold on to Grizz."

Tom then went up to the pump and reached out for the nozzle. There was a cracking sound and a small, blue flash of light between his hand and the pump. Tom jumped back and then went forward again, grabbing the fuel nozzle and directing it at Zoe. He squeezed the trigger and a clear liquid squirted out. It went all over Zoe and Grizz and, as it did so, there were a couple of loud cracks as sparks jumped away from them and into the liquid.

Zoe squealed and Grizz barked. Tom laughed loudly and fell to his knees, releasing the pump nozzle as he did so.

"It's H two O," he said, gleefully.

"Water!" said Zoe, and, before Tom could react, she rushed over and squirted the liquid over Tom for a few seconds.

The three of them sat for a moment recovering from the shock of the electrical discharge and the sudden iciness of the water.

"Water conducts electricity," said Tom. "Look, Grizz's fur is back to normal. At bit wet, but at least it's gone flat again."

"My hair's gone back down too. And so has yours," said Zoe.

"So now we're safe to refuel the jet-bug," said Tom, happily.

"What with?" said Zoe. "We can't use water, can we?"

"No, but we can use octane," said Tom. "That's the carbon eight, hydrogen eighteen in the other pump. They're not netball scores at all."

"I knew that," said Zoe. "I just wanted to get you thinking."

"I guess what happens on most platforms is that creatures go to the first pump to refuel, touch it, make a spark and then cause an explosion. All they needed to do was discharge on the water pump first."

He hauled Zoe to her feet and then went over to the other pump. Very gingerly, he touched the metal nozzle with his forefinger. To his enormous relief, there was no spark and he was able to lift the nozzle from the pump. He walked over to the jet-bug trailing the fuel line behind him. Zoe had already located the inlet to the fuel tank. It was clearly labelled 'FUEL' and she held the flap open as Tom placed the nozzle into position. He squeezed the trigger and a strong-smelling liquid gushed into the vehicle. After a minute or so the fuel tank was full and Tom replaced the nozzle onto the fuel pump.

"The jet-bug for Hitchin's Wood will depart from platform seventeen in ten seconds," blared a new announcement from the loudspeaker system.

"Quick!" shouted Tom, and they leapt inside the buggy just as the doors were closing. The jet-bug sped off out of the station and into the sunlight before any of them even had time to sit down. It was so fast that they could barely recognise anything as the scenery rushed past the windows. They sat down and simply enjoyed the ride. It didn't last all that long. After only about five minutes the jet-bug came to a halt at the next stop on the line. There was a sign displayed above the only platform of the station. It said 'Hitchin's Wood.'

The doors slid open and they stepped onto the platform. Almost immediately, the jet-bug sped off back in the direction from which they had just come.

Once again, they were alone, this time standing on a long, thin platform from which there was no obvious exit. Tom noticed a computer screen flickering. It was built into the station wall and they rushed over to look at the picture it displayed. Tom was sure that this must be the end of the Crusade.

On the screen they saw the familiar scene of Tom's bedroom and also the flashing 'PASSWORD' message. They had yet another clue letter. The screen revealed, 'M E H I C A S E D R.'

"What's the password, Tom?" said Zoe. "Surely we've finished the Crusade now."

"I don't know what it means," said Tom. "I can't think of anything that it might be."

Tom was beginning to wonder if there was yet another level to the Crusade. But before he could consider this possibility any further, he was disturbed by a white-haired man dashing along the platform. He was dressed only in a towel wrapped around his waist and draped over his shoulder, and he left wet footprints along the platform. He didn't appear to see any of them as he sped past them, a look of joy on his face.

"Eureka! Eureka!" he shouted as he ran along.

"Who or what was that?" said Zoe. "And what does 'Eureka' mean?"

"I'm not sure," said Tom. Let's follow the footprints. I think it must have something to do with the Crusade."

Chapter Eleven - The End of the Crusade

They followed the footprints for about fifty paces along the platform. They then lost the trail. The footprints simply disappeared through the wall. Upon closer inspection, Tom noticed that there was a door in the wall. It was camouflaged to look like the wall which is why he had not noticed it before. He opened the door and looked in. To his embarrassment, he saw the white-haired man was sitting in bath in the middle of a room.

"I'm terribly sorry," said Tom, blushing, and he went to close the door again.

"Eureka!" shouted the man, looking directly at Tom as though he expected some sort of reply.

Tom stood still and stared back. He was more than a little curious. The bath was absolutely full to the brim and quite a lot of water had spilled onto the wooden floor.

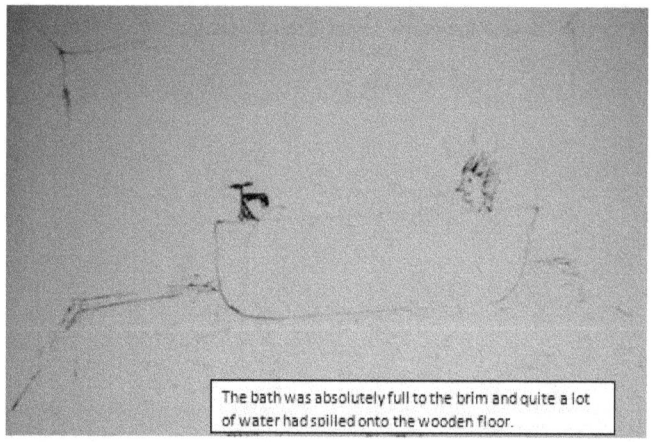

The bath was absolutely full to the brim and quite a lot of water had spilled onto the wooden floor.

"Eureka!" repeated the man, and suddenly Tom jumped upwards with excitement, punching the air.

"So have I," he said. "Thank you."

He closed the door which seemed the most polite thing to do and rushed back along the platform. Zoe and Grizz followed.

He stopped at the computer keyboard.

"The password is Archimedes," he said as Zoe approached. He was breathless with all the rushing about but also with the thrill of having solved the Crusade.

"Arky what?" said Zoe.

"Archimedes," said Tom. "It's the name of an ancient Greek scientist. He worked out the rules of floating and sinking while sitting in a bath."

'ARCHIMEDES,' he typed and then he pressed the RETURN key.

'ILLEGAL PASSWORD . . . M E H I C A S E D R,' came the reply.

"But I've found it," cried out Tom in despair. "I've found it, so why won't it work?"

He typed it again thinking that he might have made a spelling mistake the first time. He got the same reply message.

The white-haired man came rushing past them again, his arms waving jubilantly in the air.

"Eureka! Eureka!" he shouted, gleefully. And then he was gone.

"Eureka," repeated Zoe. "What does it mean?"

"It means 'I've found it'," said Tom. "When Archimedes first thought up his idea of floating he rushed round the streets telling everyone. 'Eureka' means 'I've found it'."

"Well, it's obvious, isn't it?" said Zoe.

"What is?" said Tom, confused by his sister's remark.

"Eureka is the password," she said. "How do you spell it?"

She pushed Tom out of the way and stood with her fingers poised over the keys of the computer.

"E, U, R, E, K, A," said Tom, picking up Grizz and placing his hand on Zoe's shoulder. He had a feeling that Zoe was definitely right this time.

'******' came up on the screen. Zoe pressed the RETURN key and they both looked at the screen to see what would happen. Tom could feel his heart thumping as the screen went blank and then showed a new picture.

They saw a computer graphic of the jet-bug platform at Hitchin's Wood. At the bottom of the screen was a new message.

'CRUSADE COMPLETED. YOU SCORED LEVEL TEN.'

Tom looked over his shoulder, hardly daring to believe that they were out of the Crusade at last. They were back in his bedroom. He could not contain his delight and he put his arms around Zoe's neck and hugged her.

Just then, their mother came into the room.

"Have you two just got back?" she said. "Where did you get to?"

Tom stood back from Zoe rather too quickly to go unnoticed but his mother made no comment about his sudden show of affection to his sister.

"We went to Hitchin's Wood," he said, which was not untrue. "By the way, what time is it?" he asked, innocently.

"Oh, yes," said his mother. "Here's your watch."

She took his watch out of her pocket and Tom saw that it was half past five.

"Tea will be ready in five minutes," his mother said, and she left the room, closing the door behind her.

Zoe got up and went out of the room also.

"Come on, Grizz," she said, and the little dog bounded after her.

Tom switched off the computer, went to the bathroom to wash his hands and then went downstairs. He put his jacket back into the cupboard under the stairs and then wandered into the kitchen. He noticed that his bedside lamp, complete with its crooked shade, was on the shelf next to the kitchen table.

Zoe was talking about their afternoon walk.

"And then we walked along the canal path and I suppose we got back about half an hour ago," she was saying.

Tom was confused. It wasn't like Zoe to make up something like that especially if she thought that there was a chance of getting her older brother into trouble. Perhaps the adventure hadn't happened at all. Maybe he had imagined the whole thing. Maybe his mind was playing tricks on him because he had been spending too much time staring at a computer screen. Perhaps he had dozed off while he had been sitting there and he had imagined the whole thing. Yes, that would explain it.

"Who's got Grizz's lead," said his mother, interrupting his thoughts.

"Zoe has," said Tom, immediately.

His mother turned towards him with a surprised look on her face. Tom realised that it was because he had actually called his sister by her proper name.

Zoe pulled the lead from her pocket.

"Eureka," she said, holding it up.

Tom gave her a knowing smile.